HONOUR'S ENERGY AND ENVIRONMENTAL HANDBOOK

By Walter W. Honour

a TECHNOMIC publication
TECHNOMIC Publishing Co., Inc.
265 Post Road West, Westport, CT 06880

HONOUR'S ENERGY AND ENVIRONMENTAL HANDBOOK

265 Post Road West, Westport, CT 06880

a TECHNOMIC publication

Printed in the U.S.A.
LCC·79-84754
ISBN No. 087762-274-4

ABOUT THE AUTHOR

Walter W. Honour retired from the U.S. Naval Service as Captain in 1972 after over thirty years of service in increasing roles of responsibility including a tour as Associate Professor of Naval Sciences at Purdue University, command of two squadrons, a combatant ship and a major patrol wing. On retirement, after a short time with an engineering firm, he was appointed to his present position of Chief, Bio-Environmental Services Division, Consolidated City of Jacksonville and Duval County and acts as an enforcement agent for the State Department of Environmental Regulation. His staff performs all field activities, enforcement actions and staff work for The Environmental Protection Board of the City of Jacksonville. He was appointed to a special State Task Force to develop emission standards for nitrogen oxides from existing fossil fuel fired steam generating plants. He is past State Chairman of the Florida Local Environmental Regulation Association. He holds a BS degree from the U.S. Naval Academy, and Master of Science Degrees in Administration and Engineering from George Washington University. His previous publications include numerous technical articles in Armed Forces periodicals and the books *WASHMIC* about the Washington Military Industrial Complex, and *Management of Local Pollution Control Programs.* He is listed in *WHO's WHO* in *Health* in the United States as an Environmentalist.

TABLE OF CONTENTS

PREFACE

Honour's Energy and Environmental Handbook is a ready reference of useful information compiled for use by engineers, planners, environmentalists, academicians, consultants, conservationists, and environmental enforcement personnel.

The main body of the Handbook is an alphabetical listing of terms, abbreviations, key words and phrases and their definitions commonly encountered in the following disciplines, and areas of interest: (list not totally inclusive)

- air pollution
- biology
- chemistry
- disease vector control
- electricity
- energy
- environmental terminology
- federal government terms
- geology
- hydrology
- industrial processes
- land planning
- meterology
- modelling
- noise and vibration
- petrochemical processes
- physical terms and constants
- pollution control equipment
- power generation
- radiology
- refinery operations
- sanitation
- societies and associations
- solar energy
- solid wastes control
- thermodynamics
- toxicology
- water conservation
- water pollution
- zoology

The second part of the book is devoted to tables and monographs of specific information related to the Handbook subject matter. Included are conversion tables and charts related to the environment and energy and short explanations of such topics as time, tides and currents and chart reading.

Appreciation is extended to the staff members of the Bio-Environmental Services Division for their comments and review of material to be included in the Handbook. Special recognition is accorded Ms. Corrine Harris for her painstaking clerical support.

Jacksonville, Florida
October, 1978 **Walter W. Honour**

A

A. See Argon; also ampere.

AASHO. American Association of State Highway Officials.

Ab. See Alabamine.

Abatement. The method of reducing the degree or intensity of pollution, also the use of such a method.

Abs. Absolute.

Absorber. A device utilized to extract selectively one or more elements of a gas stream from others by absorption in a liquid medium. Usually the process is performed in cylindrical towers packed with an inert material thus providing a large surface area for intimate contact between the rising gas and the falling liquid. (The process may also be carried out in a tower containing perforated trays in which the rising gas bubbles through the layer of liquid on the trays).

Absorber Plate. That component of a solar collector which absorbs the sun's radiation. In most collectors it is the metal plate to which the tubes carrying the fluid are attached or integral.

Absorptance. A property of a solar collecting surface. It is defined as the portion of the radiation that is absorbed by the surface. Color affects absorptance.

Absorption. The penetration of a substance into or through another. For example, in air pollution control, absorption is the dissolving of a soluble gas, present in an emission, in a liquid which can be extracted.

Absorption Coefficient. Volume of a gas which is dissolved at a specific temperature in 1 milliliter of liquid.

Absorption Type Air Conditioner. An air conditioning device that utilizes heat rather than electricity as its primary power source for driving the refrigeration unit.

AC. Alto cumulus clouds; alternating current.

Ac. See Actinium.

Accelerator. In radiology, a device for imparting high velocity to charged particles such as electrons or protons. These fast particles can penetrate matter and are known as radiation.

Acclimation. The physiological and behavioral adjustments of an organism to changes in its immediate environment.

Acclimatization. The acclimation or adaptation of a particular species over several generations to a marked change in the environment.

Accretion. The gradual building up of a channel bottom due to silting or wave action.

Accumulator. A vessel for the temporary storage of a gas or liquid; usually used for collecting sufficient materials for a continuous charge to a refining process.

ACE. Atmospheric Control Experimentation.

ACFM. Actual cubic feet per minute.

Acid Mist. Means any size liquid drops of any acid including but not limited to sulfuric acid and sulfur trioxide, hydrochloric acid and nitric acid as measured by approved test methods.

Acidotrophy. Water in which the concentration of hydrogen ions is high, producing highly acidic reactions.

Acid Sludge. The residue left after treating petroleum oil with sulfuric acid for the removal of impurities. It is a black, viscous substance containing the spent acid and impurities which have been separated from the oil.

Acid Treatment. An oil-refining process in which unfinished petroleum products, such as gasoline, kerosene, diesel fuel and lubricating stocks, are contacted with sulfuric acid to improve color, odor, and other properties.

Acidulate. To make acid, especially slightly acid; to treat with acid.

Acre-Foot. A term used in measuring the volume of water, equal to the quantity of water required to cover 1 acre 1 foot in depth or 43,560 cubic feet.

ACS. American Chemical Society.

ACSA. Acoustical Society of America.

Actinium (Ac). A radioactive element whose atomic weight is 227 and atomic number is 89.

Actinometer. An instrument for measuring the direct heating power of the sun's rays or the actinic power of radiant energy or for determining photographic exposure to be allowed.

Activated Carbon. A highly absorbent form of carbon, used to remove odors and toxic substances from gaseous emissions. In advanced waste treatment, activated carbon is used to remove dissolved organic matter from waste water.

Activated Sludge. Sludge that has been aerated and subjected to bacterial action, used to remove organic matter from sewage.

Activated Sludge Process. The process of using biologically active sewage sludge to hasten breakdown of organic matter in raw sewage during secondary waste treatment.

Activated Water. A transient, chemically very reactive state created in water by absorbed ionizing radiation. Water used as a coolant in a nuclear reactor becomes so activated.

Activation. The generation, under aerobic conditions of zoogleal organisms capable of absorbing organic material from sewage in the activated sludge process.

Actual Cubic Feet Per Minute. Actual or measured volume of flow at system temperature and pressure.

Acute. Having a sudden onset and a short and relatively severe course, as an acute illness.

Acute Toxicity. Any poisonous effect produced within a short period of time usually up to 24–96 hours, resulting in severe biological harm and often death.

Ad. A particular strain of laboratory mouse.

Adaptation. A change in structure or habit of an organism that produces better adjustment to the environment.

Addition Reaction. Direct chemical combination of two or more substances to form a single product, such as the union of ethylene and chlorine to form ethylene dichloride:

$$C_2H_4 + Cl_2 \longrightarrow C_2H_4Cl_2$$

Adiabatic Lapse Rate. The rate at which air lifted adiabatically cools owing to the drop of pressure with increasing height, 5.4°F/1000 ft. in the earth's atmosphere.

Adiabatic Process. A thermodynamic change of state of a system in which there is no transfer of heat or mass across the boundaries of the system.

Adiabatic Temperature. (Combustion) The theoretical temperature that would be attained by products of combustion provided the entire chemical energy of the fuel, the sensible heat content of the fuel, and combustion air above the ambient temperature were transferred to the products of combustion. This assumes (1) that combustion is complete, (2) that there is no heat loss, (3) that there is no dissociation of the gaseous compounds formed, and (4) that inert gases play no part in the reaction.

Adjutage. A tube inserted into an orifice.

ADP. Automatic Data Processing.

Adsorbent. In addition to the adjectival meaning, the term describes any of several substances that collect gaseous pollutants. Used both for measurement and control.

Adsorption. The adhesion of a substance to the surface of a solid or liquid. Adsorption is often used to extract pollutants by causing them to be attached to such adsorbents as activated carbon or silica gel. Hydrophobic, or water-repulsing adsorbents, are used to extract oil from waterways in oil spills.

Adulterants. Chemicals or substances that by law do not belong in a food, plant, animal or pesticide formulation. Adulterated products are subject to seizure by the Food and Drug Administration.

Adulticide. A chemical used to destroy or control adult arthropods such as mosquitoes or flies.

Adulticiding. The elimination or control of adult arthropods by the use of chemicals such as BAYTEX or DIBROM for mosquitoes.

Advanced Waste Treatment. Waste water treatment beyond the secondary or biological stage that includes removal of nutrients such as phosphorus and nitrogen and a high percentage of suspended solids. Advanced waste treatment, known as tertiary treatment, is the polishing stage of waste water treatment and produces a high quality effluent.

Advection. The transport of a fluid property by the mean velocity field of the fluid.

AEAP. Attainment As Expeditiously As Possible, EPA.

Aeolian Deposit. Soil deposited by wind action.

AEPS. Automated Environmental Prediction System.

Aeration. The process of being supplied or impregnated with air. Aeration is used in waste water treatment to foster biological and chemical purification.

Aerobic. This refers to life or processes that can occur only in the presence of oxygen.

Aerochlorination. The application of compressed air and chlorine gas to sewage or wastes for the removal of grease.

Aerosol. A suspension of liquid or solid particles in the air, or a gaseous medium.

Affected Facility. With reference to a stationary source, any apparatus to which a standard is applicable.

Afterbay. The tail race of a hydroelectric plant at the outlet of the turbines. Also applied to a short stretch of stream, conduit, pond or reservoir.

Afterburner. An air pollution abatement device that removes undesirable organic gases through incineration.

Ag. See silver.

AGER. Designation for an Environmental Research ship.

Agglomeration. The coalescence of dispersed suspended matter into larger flocs or particles which settle rapidly.

Aggradation. The geologic process by means of which various parts of the surface of the earth are raised in elevation or built up by the deposition of the material eroded from other sections and transported thereto by water or wind.

Agricultural Pollution. The liquid and solid wastes from all types of farming, including runoff from pesticides, fertilizers and feedlots; erosion and dust from plowing, animal manure and carcasses and crop residues and debris. It has been estimated that agricultural pollution in the U.S. has amounted to more than 2½ billion tons per year.

AGSC. Designation for a Coastal Survey Ship.

AGU. American Geophysical Union, Washington, D.C.

A & HM. Air and hazardous materials (element of EPA Regional Offices).

AIBS. American Institute of Biological Sciences, Washington, D.C.

AICE. American Institute of Consulting Engineers.

AIEE. American Institute of Electrical Engineers.

Air. So-called pure air is a mixture of gases containing about 78 per cent nitrogen; 21 per cent oxygen; less than 1 per cent of carbon dioxide, argon, and other inert gases; and varying amounts of water vapor.

Air Atomizing Oil Burner. A burner in which oil is atomized by compressed air which is forced into and through one or more streams of oil thus breaking it into a fine spray.

Air Binding. A condition occurring in the soil as a result of the increase in the pressure of the air trapped in the interstices thereof, which reduces the rate of infiltration of water. Also, the clogging of a filter, pump, or pipe due to entrained air.

Air Cooled Condenser. An air-cooled heat exchanger to condense vapors to liquids.

Air Curtain. A method for mechanical containment of oil spills. Air is bubbled through a perforated pipe causing an upward water flow that retards the spreading of oil. Air curtains are also used as barriers to prevent fish from entering a polluted body of water.

Air Curtain Destructor. A device employing an air blower with pit incinerator. Excess oxygen and turbulence resultin apparent complete combustion, leaving no residue unburned carbon (smoke) nor odorous hydrocarbons. The device has been satisfactorily demonstrated for disposal of low-ash, high-Btu waste, such as trees, tree trunks, brush (but not leaves), and wooden crating material. Excessive pollution results when materials such as automobile tires, cushions, and other non-wood wastes are burned.

Air Heater or Air Preheater. Heat transfer apparatus through which combustion air is heated by a medium of higher temperature, such as the products of combustion of steam.

Air Mass. A widespread body of air with properties that were established while the air was situated over a particular region of the earth's surface and that undergoes specific modifications while in transit away from that region.

Air Monitoring. See monitoring.

Air Ploughing. Aeration, the pumping of air into the lower oxygen, depleted layer of a water body to encourage the mixing of deep and surface waters and/or oxidation of bottom sediments.

Air Pollutant. Any matter found in the atmosphere other than oxygen, nitrogen, water vapor, carbon dioxide and the inert gases in natural concentrations.

Air Pollution. The presence of contaminants in the air in concentrations that prevent the normal dispersive ability of the air and that interfere directly or indirectly with man's health, safety or comfort or with the full use and enjoyment of his property.

Air Pollution Episode. The occurrence of abnormally high concentrations of air pollutants usually due to low winds and temperature inversion and accompanied by the increase in illness and death. See inversion.

Air Pollution Source. Any source at, from, or by reasons of which there is emitted into the atmosphere any air pollutant(s).

Air Quality. Ambient pollutant concentrations and their spatial and temporal distributions.

Air Quality Control Region. An area designated by the Federal government where two or more communities – either in the same or different states – share a common air pollution problem.

Air Quality Criteria. The levels of pollution and lengths of exposure at which adverse effects on health and welfare occur.

Air Quality Standards. The prescribed level of pollutants in the outside air that cannot be exceeded legally during a specified time in a specified geographical area.

Airshed. A term, now little used, denoting a geographical area the whole of which, because of topography, meterology, and climate, shares the same air.

Airway Resistance. The narrowing of the air passages of the respiratory system in response to the presence of irritating substances.

AISI. American Iron and Steel Institute.

Al. See aluminum.

Alabamine (Ab). An element whose atomic weight is 221 and atomic number is 85.

Albedo. Reflective power; specifically the fraction of incident, light or electromagnetic radiation that is reflected by a surface or body (as the moon or a cloud).

Algae. Primitive plants, one or many-celled, usually aquatic and capable of elaborating their foodstuffs by photo-synthesis.

Algal Bloom. A proliferation of living algae on the surface of lakes, streams or ponds. Algal blooms are stimulated by phosphate enrichment.

Algicide. Any material, substance, or compound which is fatal to algae or inhibits their growth to such an extent as to be considered a positive means of control.

Algorithm. A specific mathematical calculation procedure.

Alkali. Certain soluble salts, principally of sodium, potassium, magnesium and calcium that occur in water or soils.

Alkaline. Water or soils which contain sufficient alkali substances to raise the pH-value above 7.0 or be harmful to crop growth.

Alkalinity. A term used to represent the content of carbonates, bicarbonates, hydroxides, and occasionally borates, silicates, and phosphates in water expressed in ppm of calcium carbonate.

Alkylation. In petroleum refining, usually the union of an olefin (ethylene through pentene) with isobutane to yield high-octane, branched-chain paraffinic hydrocarbons, Alkylation may be accomplished by thermal and catalytic reactions. Alkylation of benzene and other aromatics with olefins yields alkyl aromatics.

Allochthonous. Substances in a lake that originate from sources outside the lake.

Allotrope. One of the two or more forms of an alement differing in either or both physical and chemical properties. Ozone (O_3) and molecular oxygen (O_2) are allotropes of the element oxygen (O).

Alluvial. A descriptive term referring to soil or earth material which has been deposited by water.

Alluviation. The process of accumulating deposits of gravel, sand, silt or clay, at places in rivers, lakes or estuaries, where the flow is checked.

Alpha Particle. A positively charged particle emitted by certain radioactive materials. It is the least penetrating of the three common types of radiation (alpha, beta and gamma) and usually not dangerous to plants, animals or man.

Alpha Ray. A stream of positively charged particles identified as helium nuclei – each containing two protons and two neutrons; capable of dense ionization. Often written a-ray.

Alternative Method. Any method of sampling and analyzing for an air pollutant which is not a reference or equivalent method but which has been demonstrated to the administrators (EPA) satisfaction, in specific cases, to produce results adequate for his determination of compliance.

Alum. A common name for aluminum sulfate.

Alumina. Aluminum oxide (Al_2O_3), an intermediate product of the production of aluminum. This oxide also occurs widely in nature as corundum.

Aluminum (Al). A white soft metal element whose atomic weight is 26.97 and atomic number is 13.

Alveoli. The tiny air spaces at the end of the terminal bronchioles of the lungs, where the exchange of oxygen and carbon dioxide takes place.

AM. Alveolar Macrophage.

Am. Americium.

AMA. American Medical Association.

Ambient Air. Any unconfined portion of the atmosphere; the outside air.

Amercium (Am). A completely man-made element. Atomic weight (most stable isotope) 243; atomic number 95.

Ammoniator. Apparatus used for applying ammonia or ammonium compounds to water.

Ammonification. Bacterial decomposition of organic nitrogen to ammonia.

Amoeba. A single-celled animal.

AMP. Adenosine monophosphate.

AMS. American Meteorological Society; Boston, Mass.

Anadromous. Type of fish that ascend rivers from the sea to spawn.

Anaerobic. Refers to life or processes that occur in the absence of oxygen.

Anechoic Chamber. A chamber used in sound test work which is virtually free of echoes and reverberations.

Anemometer. An instrument for measuring and indicating the force or speed of the wind.

Aneroid Barometer. A barometer in which the action of atmospheric pressure in bending a metallic surface is made to move a pointer and when calibrated indicates barometric pressure.

Angle of Incidence. The angle between the direction of the sun and the perpendicular to the surface on which the rays are falling. It affects:

a. The intensity of the direct radiation component normal to the the receiving surface. The smaller the angle the greater the intensity.
b. The ability of the surface to reflect transmit, or absorb the sun's rays.

Angle of Inclination. The angle at which the solar collector is elevated from the horizontal. Also called angle of tilt.

Angle of Tilt. See angle of inclination.

Annulus. Spacing between wall of hole and casing in well construction.

Anode. In aluminum production, the positively charged carbon terminal in the reduction cell or pot. Oxygen is attracted to the anode where it combines with carbon plus any impurities, such as sulfur, which may be present. The anode is consumed by this process and must be replaced periodically.

The positively charged terminal of a battery or cell.

ANSI. American National Standards Institute.

Anthracite Coal. A hard, black, lustrous coal containing a high percentage of fixed carbon and a low percentage of volatile matter. Commonly referred to as "hard coal," it is mined in the United States, mainly in eastern Pennsylvania, as well as in small quantities in other states.

Antibiosis. Biological action which inhibits putrefaction and oxidation.

Antichlors. Reagents such as sulfur dioxide and sodium bisulfite which can convert excess chlorine into an inert salt.

Anticoagulant. A chemical that interferes with blood clotting, often used as a rodenticide.

Anti-degradation Clause. A provision in air quality and water quality laws that prohibits deterioration of air or water quality in areas where the pollution levels are presently below those allowed.

Antimony (Sb). A brittle blue-white metal element whose atomic number is 51 and atomic weight is 121.76.

API. Air Pollution Institute operated by EPA at Research Triangle Park, N.C.; also, American Petroleum Institute.

APRAC-IA. Carbon Monoxide Model (CO), EPA.

APTD. Air Pollution Technical Data.

APTI. Air Pollution Training Institute (EPA).

AQ. Air Quality, EPA.

AQCR. Air Quality Control Region..

AQDHS. Air Quality Data Handling System (EPA).

AQDM. Air Quality Demonstration Model; A mathematical model whose computer output is used to portray the approximate air quality of an area under varying conditions.

AQMA. Air Quality Maintenance Areas

AQMP Air Quality Maintenance Plan

Aquatic Plants. Plants that grow in water either floating on the surface, growing up from the bottom of the body of water or growing under the surface of the water.

Aqueduct. A large conduit used to convey water.

Aqueous Metamorphism. Changes in the texture and composition of rocks brought about by solution and redisposition of mineral matter through the influence of water which they contain.

Aquiclude. A geologic formation which, although porous and capable of absorbing water slowly, will not transmit it rapidly enough to furnish an appreciable supply for a well or spring.

Aquifer. An underground bed or stratum of earth, gravel or porous stone that contains water.

Aquifuge. A rock which contains no interconnected openings or interstices and therefore neither absorbs nor transmits water.

Area Source. A category of emitters that are individually minor but sufficiently numerous and widespread that their combined emissions are significant, e.g. automobiles.

Argon (A). An inert gaseous element present in the atmosphere with atomic weight 39.94 and atomic number 18.

Arid. An area where precipitation is insufficeint to sustain agriculture without irrigation.

Arroyo. A water-carved channel in arid areas with small steep banks which is dry except for infrequent rainfall when it becomes a creek or river.

ARS. Agriculture Research Service (USDA).

Arsenic (AS). A steel-gray crystalline, brittle solid element whose atomic weight is 74.91 and atomic number is 33. The element is non-poisonous but many of its compounds are extremely poisonous.

Artesian. Groundwater under pressure which will rise to a higher elevation if afforded the opportunity to do so.

As. See Arsenic.

ASAC. Altostratus and Altocumulus.

ASAP. As Soon As Possible, EPA.

Asbestos. A mineral fiber with countless industrial uses; a hazardous air pollutant when inhaled.

A-Scale Sound Level. The measurement of sound approximating the auditory sensitivity of the human ear. The A-Scale sound level is used to measure the relative noisiness or annoyance of common sounds.

ASCE. American Society of Chemical Engineering.

Ash. The noncombustible solid matter in fuel.

Ash-Free Basis. The method of reporting fuel analysis whereby ash is deducted and other constituents are recalculated to total 100 per cent.

ASII. American Science Information Institute; Detroit, Michigan.

ASIRC. Aquatic Sciences Information Retrieval Center; Kingston, Rhode Island.

ASME. The American Society of Mechanical Engineers.

ASO. American Society for Oceanography.

Aspirating Burner. A burner in which the fuel in a gaseous or finely divided form is burned in suspension. The air of combustion is supplied by drawing it through one or more openings by the lower static pressure created by the velocity of the fuel stream.

Assimilation. Conversion or incorporation of absorbed nutrients into protoplasm. Also refers to the ability of a body of water to purify itself of organic pollution.

ASTM. American Society of Testing Materials.

A-strain. A particular type of influenza virus.

ATM. Atmospheric Transport Model, EPA.

Atm. Atmosphere.

ATMI. American Textile Manufacturers Institute.

Atmometers. Instruments used to study evaporation from plants by having special moist surfaces from which water loss is recorded.

Atmosphere. The layer of air surrounding the earth.

Atmosphere, Length of. A length expressed in terms of airmass, m, which is the ratio of the total atmosphere in the actual earth-sun path to the mass which would exist if the sun were directly overhead at sea level where m=1.0 (solar energy term).

Atmospheric Area. As the federal government uses the term, a segment of the continental United States in which climate, meterology, and topography – all of which influence the capacity of the air to dilute and disperse pollutants – are essentially similar.

Atmospheric Pressure. The pressure due to the weight of the atmosphere. Normal atmospheric pressure at sea level is approximately 14.7 p.s.i. or 29.92 inches of mercury.

Atom. The smallest unit of an element to retain the chemical characteristics of that element. It consists of negatively charged particles called electrons orbiting around the nucleus and, within the small mass of the nucleus, other particles protons (which are positively charged and balance the extranuclear electrons) and neutrons (which have no charge). It is the number and arrangement of the atom's electrons and protons that make one element differ from another; the number of neutrons distinguishes one isotope from another of the same element.

Atomic Number. For each element, the total number of protons in its nucleus. It is also the total number of electrons. See atom.

Atomic Pile. A nuclear reactor.

Atomic Weight. The average relative weight of an atom of an element as compared with another element (usually oxygen) that is taken as a standard. Isotopes of the same element have different atomic weights because the number of neutrons differs from isotope to isotope. a quoted atomic weight is the average weight unless a specific isotope is named.

Atomizer. A device by means of which a liquid is reduced to a very fine spray.

Attractant. A chemical or agent that lures insects or other pests by olfactory stimulation.

Attrition. Wearing or grinding down by friction. One of the three basic contributing processes of air pollution, the others are vaporization and combustion.

AtWt. Atomic Weight.

Audiometer. An instrument for measuring hearing sensitivity.

Autochthonous. Substances in a lake that originate from sources in a lake.

Autotrophic. Self-nourishing; denoting those organisms capable of constructing organic matter from inorganic substances.

Available Heat. The quantity of useful heat per unit of fuel available from complete combustion after deducting dry flue gas and water vapor losses.

Avg. Average

AVGAS. Aviation gasoline

AVOIL. Aviation oil.

Avulsion. The process in which a stream suddenly breaks through its confining banks and forms another unexpected channel – also rapid erosion of shoreland by wave action.

AWS. Average Wind Speed.

B

B. See Boron.

Ba. See Barium.

Backfill. The material used to refill a ditch or other excavation, or the process of doing so.

Backflow Preventer. A device or check valve in plumbing lines which restricts the flow of liquids to one desired direction only. Usually required in all cross connections to public utility water and sewer lines.

Background Concentration. See background level.

Background Level. With respect to air pollution, amounts of pollutants present in the ambient air due to natural sources. Background levels may also include pollutants advected into the area.

Background Radiation. Normal radiation present in the lower atmosphere from cosmic rays and from earth sources.

Backing. A wind that is shifting in its source direction counter clockwise.

Backwashing. The operation of cleaning a filter or plant flow through section by reversed flow of water or other liquids.

BACT. Best Available Control Technology.

Bacteria. Single-celled microorganisms that lack chlorophyll. Some bacteria are capable of causing human, animal or plant diseases, others are essential in pollution control because they breakdown organic matter in the air and in the water.

Baffle. Any deflector device used to change the direction of flow or the velocity of water, sewage or products of combustion such as fly ash or coarse particulate matter. Also used in deadening sound.

Bagasse. Sugar cane from which the juice has been essentially extracted.

Bag Filter. A device containing one or more cloth bags for recovering particles from the dust-laden gas which is blown through it.

Baghouse. An air pollution abatement device used to trap particulates by filtering gas streams through large fabric bags, usually made of glass fibers.

Bag-type Collector. A filter wherein the cloth filtering medium is made in the form of cylindrical bags.

Baling. A means of reducing the volume of solid waste by compaction.

Ballistic Separator. A machine that separates inorganic from organic matter in a composting process.

Band Application. With respect to pesticides, the application of the chemical over or next to each row of plants in a field.

Banking. Burning solid fuels on a grate at rates sufficient to maintain ignition only.

Banquette. A berm formed at the landward base of a levee to increase its strength.

Bar. An alluvial deposit at the mouth of a stream or estuary. Also any deposit in a stream causing an obstruction to navigation.

Barium (Ba). A soft silvery white metallic element whose atomic weight is 137.36 and atomic number is 56.

Bark Boiler. A combustion unit designed to burn mainly bark and wood residues, used to produce steam for process or electrical energy.

Barometric Condenser. An inexpensive direct contact condenser used when condensate recovery is not a factor. In this type of condenser, steam rises into a rain of cooling water, and both condensed steam and water flow out of the bottom of the condenser, maintaining a partial vacuum in the condenser.

Barothermograph. An instrument which records simultaneous barometric pressure and temperature on the same chart.

Barrier. A dam or sill in a stream to constrain its flow. Also, a buffer zone for noise reduction.

Bar Screen. In waste water treatment, a screen that removes large floating and suspended solids.

Basal Application. With respect to pesticides, the application of the pesticide formulation on stems or trunks of plants just above the soil line.

Base Stock. A sheet, usually produced from unbleached kraft pulp, formed into linerboard on a fourdrinier machine.

Basin. See river basin.

BAT. Best Available Technology.

Batch Fed Incinerator. An incinerator that is charged with refuse periodically the charge being allowed to burn down or burn out before another charge is added.

Batture. An elevated river bed.

BC Chart. Bottom contour chart.

BD. Weather abbreviation for "blowing dust."

Be. See Beryllium.

Beaufort Wind Scale. A scale of wind force from 0–17 where force 0 is calm and forces 12–17 are hurricane winds of more than 73 mph. e.g. Force 5 is fresh breeze, 19–24 mph.

Bellmouth. A rounded enlarged entrance to a pipe or orifice.

Benthic Region. The bottom of a body of water. This region supports the benthos, a type of life that not only lives upon, but contributes to the character of the bottom.

Benthos. The plant and animal life whose habitat is the bottom of a sea, lake or river.

Berkelium (Bk). A man-made radio-active element; second element in the curide series. Atomic weight (most stable known isotope) 247; atomic number 97.

Berm. A horizontal strip or shelf built into an embankment to increase thickness and strength against erosion.

Beryllium (Be). A hard metallic element used in alloys with atomic weight of 9.02 and atomic number 4. Sometimes called Glucinum (GL).

Beta Particle. An elementary particle emitted by radioactive decay that may cause skin burns. It is easily stopped by a thin sheet of metal.

Beta Ray. A stream of negatively charged electrons ejected from the nucleus of an atom; more penetrating than an alpha ray but producing less ionization. Often written B-ray.

BEV. Billion-electron-volts.

BHA. Butylated hydroxyanisole.

BHP. Brake horsepower.

BHT. Butylated hydroxytoluene.

Bi. See Bismuth.

Bib. A valve closed by screwing down a leather or fiber washered disk on to a seat in the valve body.

Binder. See core binder.

BINOVC. Weather abbreviation for "breaks in the overcast."

Bioassay. The employment of living organisms to determine the biological effect of some substance, factor or condition.

Biochemical Oxygen Demand (BOD). A measure of the amount of oxygen consumed in the biological processes that break down organic matter in water. Large amounts of organic waste use up large amounts of dissolved oxygen, thus the greater the degree of pollution, the greater the BOD.

Biodegradable. The process of decomposing quickly as a result of the action of microorganisms.

Biological Control. A method of controlling pests by means of introduced or naturally occurring predatory organisms, sterilization or the use of inhibiting hormones, etc., rather than by mechanical or chemical means.

Biological Half-Life. Also known as effective half-life; the time required for the body to excrete ½ the radiation dose of a specific substance. Different isotopes have different biological half-lives, not necessarily similar to their radioactive half-lives. See half-life.

Biological Oxidation. The process by which bacterial and other microorganisms feed on complex organic materials and decompose them. Self-purification of waterways and activated sludge and trickling filter waste water treatment processes depend on this principle. The process is also called biochemical oxidation.

Biological Treatment. Process in which living microorganisms are mixed with incoming waste, and use the biologically degradable organics in the waste as foodstuffs or as an energy source, thus effectively removing them from applied wastewaters. Distinguished from physical/chemical methods, which do not employ living microorganisms. Activated sludge and trickling filters are the two prime forms of biological treatment.

Biomass. Total amount of organic matter in a waste water treatment plant (activated sludge or trickling filter) or water body per unit or unit volume.

Biomonitoring. The use of living organisms to test the suitability of effluent for discharge into receiving waters and to test the quality of such waters downstream from a discharge.

Biosphere. The portion of the earth and its atmosphere capable of supporting life.

Biostabilizer. A machine used to convert solid waste into compost by grinding and aeration.

Biota. All the species of plants and animals occurring within a certain area.

Bismuth (Bi). A white crystalline brittle metal element used as a special alloy for castings and in some medicines. It has an atomic weight of 209.00 and atomic number 83.

Bituminous Coal. Soft coal, dark brown to black in color, having a relatively high proportion of gaseous constituents and usually burning with a smoky luminous flame.

Bivane. Bi-directional wind vanes which indicate the direction from which the wind originates in both horizontal and vertical directions.

Bk. Berkelium.

Black Liquor. Spent chemical solution which is formed during the cooking of wood pulp in the digester. The black liquor is burned as a fuel in the recovery furnace.

Blast Furnace. A shaft furnace in which solid fuel is burned with an air blast to smelt ore.

Bleeder. A bypass or relief valve used to relieve excess pressure.

Blister Copper. An impure intermediate product in the refining of copper, produced by blowing copper bearing material in a converter; the name is derived from the large blisters on the cast surface that result from the liberation of SO_2 and other gases.

Bloom. A proliferation of living algae and/or other aquatic plants on the surface of lakes or ponds. Blooms are frequently stimulated by phosphate enrichment.

Blowback. The difference between the pressure at which a safety valve opens and at which it closes, usually about three percent of the pressure at which the valve opens.

Blowdown. Hydrocarbons purged during refinery shutdowns and startups which are manifolded for recovery, safe venting or flaring.

BMP. Best Management Practice – EPA.

BOD. Biochemical Oxygen Demand.

BOD_5. The amount of dissolved oxygen consumed in five days by biological processes breaking down organic matter in an effluent. See biochemical oxygen demand.

Bog. Wet, spongy land usually poorly drained, highly acid and rich in plant residue.

Boiler. A closed pressure vessel in which the liquid, usually water, is vaporized by the application of heat.

Boiler Horsepower. A unit of rate of water evaporation. One boiler horsepower equals the evaporation of 34.5 lb. of water per hour from a temperature of 212°F into dry saturated steam at the same temperature (equivalent to 33,472 Btu per hour).

BOM. Bureau of Mines (USDI).

Bond or Chemical Bond. See valence.

Boom. A floating device that is used to contain oil on a body of water.

Boron (B). A mineral element whose atomic weight is 10.82 and atomic number is 5. Important compounds are boric acid and borax.

Borrow Pit. An excavated area usually filled with water table water in low lying coastal areas.

Botanical Pesticide. A plant-produced chemical used to control pests; for example, nicotine, strychnine or orpyrethrun.

BP. Blood pressure.

BPT. Best practicable technology.

Br. See Bromine.

Brackish Water. A mixture of fresh and salt water, or water that has excess chloride to the extent that it is unpalatable.

Brasses. Copper-based alloy of 60–65% copper. Alloying material is usually zinc.

Breaker. In anthracite mining, the structure in which the coal is broken, sized, and cleaned for market. Also known is a coal breaker. A machine used for the primary reduction of coal, ore or rock.

Breeching. A sheet-iron or sheet-metal casing at the end of boilers for conveying the smoke from the flues to the smokestack.

Breeder. A nuclear reactor that produces more fuel than it consumes.

Brightening. The process of producing bright stock (see bright stock).

Bright Stock. Refined high viscosity lubricating oils usually made from residual stocks by suitable treatment, such as a combination of acid treatment or solvent extraction with dewaxing or clay finishing.

British Thermal Unit (Btu). The mean British thermal unit is 1/180 of the heat required to raise the temperature of one pound of water from 32°F to 212°F at a constant atmospheric pressure. It is about equal to the quantity of heat required to raise one pound of water 1°F. A Btu is essentially 252 calories.

Broadcast Application. With respect to pesticides, the application of a chemical over an entire field, lawn or other area.

Bromine (Br). A heavy, mobile reddish brown non-metallic liquid element with an atomic weight 79.916 and atomic number 35. It's vapors are disagreeably odorous. It is used in photography and medicines.

Bronchial Asthma. Abnormal responsiveness of the air passages to certain substances. An attack consists of a widespread narrowing of the bronichioles by muscle spasm, swelling of the mucous membrane, or thickening and increase of mucous secretions, accompanied by wheezing, gasping, and sometimes coughing.

Bronchiole. Small branch of the bronchus.

Bronchus. A major airway of the respiratory system.

Bronzes. Copper based alloy of 85–90% copper. Alloying material is usually tin.

Brownian Motion. The random movement of microscopic particles suspended in a fluid medium.

B-Scale. Noise frequency scale.

BSFW. Bureau of SportFisheries and Wildlife; U.S. Department of the Interior.

Btu. British Thermal Unit.

Bulkhead Line. The line formed at the shore of a body of water of man-man structures to reduce avulsion and define the shoreline.

BUMINES. Bureau of Mines; U.S. Department of the Interior.

Bunker C Oil. Residual fuel oil of high viscosity commonly used in marine and stationary power plants (No. 6 fuel oil).

Buoyant Plume. A plume initially of lower density than the ambient fluid after the pressure is adiabatically brought to equilibrium. Usually, the term "buoyant plume" refers to a plume in which the effect of the initial momentum is small, and the term "forced plume" refers to a plume with buoyancy in which the effect of the initial momentum is also important.

Burial Ground (graveyard). A place for burying unwanted radioactive materials to prevent radiation escape, the earth or water acting as a shield. Such materials must be placed in water-tight, noncorrodible containers so the radioactive material cannot leach out and invade underground water supplies.

Burner. A device for the introduction of fuel and air into a furnace at the desired velocities, turbulence, and concentration to establish and maintain proper ignition and combustion of the fuel.

Buss (Busbar). A heavy metal conductor, usually copper, for high amperage electricity.

Bustle Pipe. In steel making, a metal tube of large diameter which surrounds a blast furnace at a level a little above the tuyeres; it is lined with refractory material and distributes the hot air from the blast stoves to the pipes (goosenecks) which carry the air to the tuyeres.

By-Channel. A water bypass around the end of a reservoir or dam to convey flood waters.

C

C. See Carbon.

C°. Temperature Centigrade or Celsius degrees.

Ca. See calcium.

CAA. Clean Air Act.

Cadmium (Cd). A soft blue-white metallic element with an atomic weight of 112.41 and atomic number of 48. It is used in metal alloys.

CAFB. Chemically Active Fluid Bed.

Caisson. A chamber, usually sunk by interior excavation for the purpose of gaining access to the bed of a body of water. It is frequently used as a form to pour subsurface concrete as a footing for a structure near or on a body of water.

Cake Sludge. Air dried or dewatered sludge which can be shoveled.

Cal. Calorie.

Calcine. Ore or concentrate which has been treated by calcination or roasting and which is ready for smelting.

Calcining. Roasting of ore in an oxidizing atmosphere usually to expel sulfur or carbon dioxide. If sulfur removal is carried to practical completion, the operation is termed "sweet roasting;" If all CO_2 is removed, the operation is termed "dead roasting."

Calcium (Ca). An alkaline earth element with an atomic weight of 40.08, an atomic number of 20.

Calibration. The adjustment of instrumentation so that visual readings are as accurate as possible across numerical span limits. Also, an adjustment applied to concentration estimates, based on a comparison with measured air quality data, in order to improve the accuracy of an air model.

Caliche. A hard deposit consisting mostly of calcium carbonate found in subsoil of arid areas.

Californium (CF). Man-made radio-active element. Atomic weight (most stable known isotope) 251; atomic number 98.

Calorie. The mean calorie is 1/100 of the heat required to raise the temperature of one gram of water from 0°C to 100°C at a constant atmospheric pressure. It is about equal to the quantity of heat required to raise one gram of water 1°C.

Cancer. An abnormal, potentially unlimited, disorderly new cell growth.

Capillarity. The degree to which a material with vertical pores will draw the surface of a liquid above hydrostatic pressure level when immersed in a liquid.

Carbon (c). A common solid element which forms the base for an infinite number of compounds. Atomic weight is 12.00 and atomic number is 6.

Carbonaceous Fuel. Means solid materials composed primarily of vegatative matter such as tree bark, wood waste, bagasse, and/or the combustible fraction of municipal waste.

Carbonaceous Fuel Burning Equipment. A fire box, furnace or combustion device which burns carbonaceous fuel or a combination of carbonaceous and fossil fuel for the primary purpose of producing thermal energy which is used indirectly to produce steam or to heat other liquids or gases.

Carbon Dioxide (CO_2). A colorless, odorless, nonpoisonous gas that is a normal part of the ambient air. CO_2 is a product of fossil fuel combustion, and some researchers have theorized that excess CO_2 raises atmospheric temperatures.

Carbonization. The process of converting coal to carbon in the absence of air by using intense heat to remove volative ingredients.

Carbon Loss. The loss representing the unliberated thermal energy caused by failure to oxidize some of the carbon in the fuel.

Carbon Monoxide (CO). A colorless, odorless, highly toxic gas that is a normal byproduct of incomplete fossil fuel combustion. CO, one of the major air pollutants, can be harmful in small amounts if breathed over a certain period of time.

Carbon/Nitrogen Ratio. An expression of the relative amounts of these two elements. Useful in classification.

Carburetor. A device for supplying certain internal-combustion engines with a mixture of vaporized fuel and air.

Carcinogenic. Cancer producing.

Carryover. The chemical solids and liquid entrained in the stream from a boiler or effluent from a fractionating column, absorber, or reaction vessel. Also, the conveyance of uncombusted fuels or carbon into the exhausts from internal combustion engines.

Cascade. A stretch of stream where the drop in elevation of the stream bed is considerable but not enough to cause the water to fall vertically. Descriptively a cascade is between a rapids and a waterfall.

Cascade Impactor. See Impactor.

Casing. A pipe installed in a well to prevent caving and/or circulation of water into other acquifers.

Catalase. An enzyme that catalyzes decomposition of hydrogen peroxide into water and gaseous oxygen.

Catalysis. The facilitation of a chemical reaction induced by a material – called the catalyst – that remains unchanged in the process.

Catalyst. A substance which initiates or accelerates a chemical reaction without apparent change in its own physical or chemical properties.

Catalytic Converter. An air pollution abatement device that removes organic contaminants by oxidizing them into carbon dioxide and water through chemical reaction. Can be used to reduce nitrogen oxide emissions from motor vehicles.

Catalytic Cracking. The conversion of high boiling hydrocarbons into lower boiling substances by means of a catalyst which may be used in a fixed bed, moving bed or fluid bed. Natural or synthetic catalysts are employed in bead, pellet, or powder form. Feedstocks may range from naphtha cuts to reduced crude oils.

Catch Basin. A subsurface water basin, chamber, or well usually built at the curb line of the street for the admission of surface waters and usually designed with a grit detention and removal feature.

Cathode. In aluminum production, the negatively charged terminal of the reduction cell to which the aluminum migrates. The terminal consists of the carbon lining that makes up the bottom of the cell.

Caustic Soda. Sodium hydroxide (NaOH), a strongly alkaline, caustic substance used as the cleaning agent in some detergents.

Caving. In metal mining, caving implies the dropping of the over-burden as part of the system of mining.

Cavitation. The formation of a void, or series of voids, within the body of a liquid caused by the inability of the particles of the fluid to follow normal paths because of external dynamics of pressure or mechanical agitation.

CB. Cumulo-nimbus cloud formation; citizens band radio.

Cb. See Columbium.

CBD. Central Business District.

CBO. Corrected by owner; a term used when the owner of a facility has corrected the cause of an environmental violation.

CBR. Chemical, biological, radiological.

CC. Cirrocumulus cloud formation.

Cd. See Cadmium.

CDD. See Cooling Degree Days.

CDE. Compliance Date Extension (EPA).

CDHS. Comprehensive Data Handling System (EPA).

CDM. Climatological Dispersion Model, EPA.

CDMQC. Climatological Dispersion Model Quality Control (Model for TSP, SO_2, NO_x, CO), EPA.

CdS. Cadmium sulfide.

CDS. Compliance data system, EPA.

Ce. See Cerium.

Cells. With respect to solid waste disposal, earthen compartments in which solid wastes are dumped, compacted and covered over daily with layers of earth.

Celsius. See centigrade.

Cement Grout. A mixture of cement, sand, and water (usually 2 parts of sand to one part cement and 5–7 gallons of water per bag (94 lbs.) of cement) used in well construction or modification.

Centigrade. A thermometric scale with 0° the freezing point of water and 100° the boiling point of water.

Centrifugal Collector. Any of several mechanical systems using centrifugal force to remove aerosols from a gas stream.

Centrifuge. A mechanical device utilizing centrifugal force to separate solids from liquids or for separating liquid emulsions.

CEQ. Council on Environmental Quality; Executive Office of the President, Washington, D.C.

Cerium (Ce). A steel gray lustrous metal element whose atomic weight is 140.13 and atomic number is 58. Used in incandescent gas mantles and as a pyrophoric alloy.

Cesium (Cs). An alkali metal element whose atomic weight is 132.91 and atomic number is 55. Used in radio tubes and as a catalyst in the hydrogenation of certain organic compounds.

Cesspool. An underground pit into which raw household sewage or other untreated liquid waste is discharged and from which the liquid seeps into the surrounding soil or is otherwise removed.

Cf. Californium.

CFR. Code of Federal Regulations.

cfs. Cubic feet per second, a measure of the amount of water passing a given point.

CGS. Centimeter – gram – second system.

Channelization. The straightening and deepening of streams to permit water to move faster, to reduce flooding or to drain marshy acreage for farming. However, channelization reduces the organic waste assimilation capacity of the stream and may disturb fish breeding and destroy the stream's natural beauty.

Charco. A ground surface depression either natural or artificial that impounds water for soil conservation.

Charging. Feeding raw material into an apparatus, for example, into a furnace, for treatment or conversion.

CHE. Cholinesterase.

Check. A structure or device designed to impede or control the flow of fluids.

Chemical Bond. See valence.

Chemical Energy. The energy contained in the chemical bond between atoms; it can be released to the environment by a chemical reaction; e.g. combustion.

Chemical Oxygen Demand (COD). A measure of the amount of oxygen required to oxidize organic and oxidizable inorganic compounds in water. The COD test, like the BOD test, is used to determine the degree of pollution in an effluent.

Chemosterilant. A pesticide chemical that controls pests by destroying their ability to reproduce.

Chilling Effect. The lowering of the earth's temperature due to the increase of atmospheric particulates that inhibit penetration of the sun's energy.

Chinook. A warm dry wind which blows down the slopes of mountain ranges in the western parts of the United States.

Chlorinated Hydrocarbons. A class of generally long-lasting broad-spectrum insecticides of which the best known is DDT, first used for insect control during World War II. Other similar compounds include aldrin, dieldrin, heptachlor, chlordane, lindane, endrin, mirex, benzene hexachloride (BHC), and toxaphene. The qualities of persistence and effectiveness against a wide variety of insect pests were long regarded as highly desirable in agriculture, public health and home uses. But later research has revealed that these same qualities may represent a potential hazard through accumulation in the food chain and persistence in the environment.

Chlorination. The application of chlorine to drinking water, sewage or industrial waste for disinfection or oxidation of undesirable compounds.

Chlorinator. A device for adding a chlorine-containing gas or liquid to drinking or waste water.

Chlorine (CL). A halogenic gaseous element with reddish brown color and odorous vapors. It has an atomic weight of 35.457 and an atomic number of 17. It is found in nature in combined state as common salt (NaCl).

Chlorine-Contact Chamber. In a waste treatment plant, a chamber in which effluent is disinfected by chlorine before it is discharged to the receiving waters.

Chlorine Demand. The difference between the amount of chlorine added to water sewage or industrial wastes and the amount of residual chlorine remaining at the end of a specified contact period. It varys with the amount, time of contact and temperature.

Chlorine Residual. The total amount of chlorine remaining in water, sewage, or industrial wastes at the end of a specified contact period following chlorination.

Chlorosis. Yellowing or whitening of normally green plant parts. It can be caused by disease organisms, lack of oxygen or nutrients in the soil or by various air pollutants.

Chromium (CR). A metallic element resembling iron with an atomic weight of 52.01 and an atomic number of 24. It is used widely in alloys and plating processes as a finish material.

Chromosome. A part of all cells of all organisms. Chromosomes are composed of genes and transmit characteristics from generation to generation.

Chronic. Marked by long duration or frequent recurrence as a disease.

CI. Cirrus cloud formation.

Cienega. An area where the water table is at or near the surface of the ground. Standing water occurs in depressions and other parts have grasses and heavy vegetation.

CIIT. Chemical Industry Institute of Toxicology.

Cilia. Hairlike cells that line the airways and by their sweeping movement propel the dirt and germ-filled mucus out of the respiratory tract.

Cinders. Particles not ordinarily considered as fly ash or dust because of their greater size; these particles consist essentially of fused ash and/or unburned matter.

Cistern. A small covered tank for storing water for a home or farm and usually placed underground and refilled by collecting rainwater.

CL. See Chlorine.

CLA. Coal Leasing Amendments.

Clarification. In waste water treatment, the removal of turbidity and suspended solids by settling, often aided by centrifugal action and chemically induced coagulation.

Clarifier. In waste water treatment, a settling tank which mechanically removes settleable solids from wastes.

Cleaning Fires. The act of removing ashes from the fuel bed or furnace.

Climatology. The study of weather and its effects on man and his activites. Meteorology.

Clinical Study. One that concerns itself with the disease process in the living subject.

Clinkers, Cement. The glassy, stony, lump-like product of fusing together clay and limestone as the first stage in the manufacture of portland cement.

Closed System. A system in which a fluid is contained within piping and tanks.

Cm. Curium.

CMHSA. Coal Mine Health and Safety Act.

CNEL. Community Noise Equivalent Level.

CNR. Composite Noise Rating.

CNS. Central Nervous System; the brain and spinal cord.

CO. Carbon Monoxide.

Co. See Cobalt.

Coagulant. A material which when added to sewage or water will combine with certain substances ordinarily present and form a precipitate comprising floc particles more or less gelatinous in character having the capacity to remove colloids from sewage or water.

Coagulation. The clumping of particles in order to settle out impurities; often induced by chemicals such as lime or alum.

Coal Desulfurization. See desulfurization.

Coal Gas. Gas formed by the destructive distillation of coal.

Coal Tar. A black viscous liquid formed as a by-product from the distillation of coal.

Coastal Zone. Coastal waters and adjacent lands that exert a measurable influence on the uses of the sea and its ecology.

COB. Close of business.

Cobalt (Co). A grey hard very magnetic metal element whose atomic weight is 58.94 and atomic number is 27. It is used as an alloy and it's salts produce brilliant blue colors.

COD. See chemical oxygen demand.

COE. Crude oil equivalent.

Coefficient of Haze (COH). A measurement of visibility interference in the atmosphere.

Cofferdam. A temporary structure built around a site to allow the removal of water therefrom for construction purposes.

Coffin. A thick-walled container (usually lead) used for transporting radioactive materials.

COH. See coefficient of haze.

Cohesion. The force of attraction between the molecules of any substance which tends to hold them together.

Coke. Bituminous coal from which the volatile constituents have been driven off by heat so that the fixed carbon and the ash are fused together.

Coke Breeze. Fine coke particles leaving the coke quencher with the quenched coke by conveyor. The particles are very fine and may be blown away.

Coke, Petroleum. The solid carbonaceous residue remaining as the final product of the condensation processes in cracking. It consists of highly polycyclic aromatic hydrocarbons very poor in hydrogen. It is used extensively in metallurgical processes. Calcination of petroleum coke can yield almost pure carbon or artificial graphite suitable for production of electrodes, motor brushes, dry cells, etc.

Coking. 1. Carbonization of coal by destructive distillation. 2. In petroleum refining: any cracking process in which the time of cracking is so long that coke is produced as the bottom product; thermal cracking for conversion of heavy, low-grade oils into lighter products and a residue of coke; or the undesirable building up of coke or carbon deposits on refinery equipment.

Coliform Index. An index of the purity of water based on a count of its coliform bacteria.

Coliform Organism. Any of a number of organisms common to the intestinal tract of man and animals whose presence in waste water is an indicator of pollution and of potentially dangerous bacterial contamination.

Collection Efficiency. The ratio of the weight of pollutant collected to the total weight of pollutant entering the collector.

Collector. The component of a solar system which collects the solar radiation.

Collector Bearing. The compass direction which a solar collector surfaces face; a collector bearing of 180° would mean the collector is bearing due south.

Collector Efficiency. The output, (energy collected) divided by the input (solar energy falling on solar collector) within a specified period of time.

Colloid. A substance composed of extremely small particles, ranging from 0.005 micro-meters to 0.2 micro-meters, which when mixed with a liquid will not settle, but will remain suspended. The colloidal suspension thus formed has properties that are quite different from the simple solution of the two substances. In fuel burning, a finely divided organic substance which tends to inhibit the formation of dense scale and results in the deposition of sludge, or causes it to remain in suspension, so that it may be blown from the boiler.

Colloidal Fuel. Mixture of fuel oil and powdered solid fuel.

Colloidors. Surfaces or baffles hung in a sewage tank to aid in coagulating colloidal matter.

Colloids. Finely divided solids, usually considered to be less than two microns, which will not settle naturally, but may be removed by coagulation, centrifuge or bio-chemical action.

Columbium (Cb). A rare gray metallic element with an atomic weight of 92.91 and an atomic number of 41. Also called Niobium (Nb).

Combination Boiler. A combustion unit used to produce steam for process or electrical energy which is designed to burn bark and at least one other fuel.

Combined Sewers. A sewerage system that carries both sanitary sewage and storm water runoff. During dry weather, combined sewers carry all waste water to the treatment plant. During a storm, only part of the flow is intercepted because of plant overloading; the remainder goes untreated to the receiving system.

Combustion. Burning. Technically, a rapid oxidation accompanied by the release of energy in the form of heat and light. It is one of the three basic contributing factors causing air pollution, the others are attrition and vaporization.

Combustion Contaminants. Particulate matter discharged into the atmosphere from the burning of any kind of material containing carbon.

Combustion Tower. Refractory graphite-lined or water-jacketed stainless steel tower in which phosphorus is burned to phosphorus pentoxide.

Comminutor. A device that grinds solids to make them easier to treat.

Community Air. The outside air shared by a community, rather than the perhaps particularly polluted air in the immediate vicinity of a factory.

Compaction. Reducing the bulk of solid waste by rolling and tamping.

Compliance Schedule. A legally enforceable schedule specifying a date or dates by which a source or category of sources must comply with specific emission or effluent standards contained in a plan or with the increments of progress to achieve such compliance.

Compost. Relatively stable decomposed organic material.

Composting. A controlled process of degrading organic matter by microorganisms. (1) mechanical – a method in which the compost is continuously and mechanically mixed and aerated. (2) ventilated cell-compost is mixed and aerated by being dropped through a vertical series of ventilated cells. (3) windrow – an open-air method in which compostable material is placed in windrows, piles or ventilated bins or pits and occasionally turned or mixed. The process may be anaerobic or aerobic.

Computer Code. A set of statements that comprise a computer program.

Concrete Grout. A mixture of sand, gravel, cement and water (usually one bag [96 lb.] cement, an equal volume of sand and gravel mixture and 5–7 gallons of water) used in well construction or modification.

Condensation. The change of a fluid from gas to liquid accompanied by a reduction in volume and temperature.

Condensed Fumes. Minute solid particles generated by the condensation of vapors from solid matter after volatilization from the molten state, or generated by sublimation, distillation, calcination, or chemical reaction when these processes create airborne particles.

Condenser Boiler. A boiler in which steam is generated by the condensation of a vapor.

Conduction. The transfer of heat by physical contact between substances.

Conifer. A tree or shrub that bears cones, as pines and firs.

Coning. A term describing plume rise that is evenly diffused vertically and horizontally resulting in a conical shape with apex at source. Occurs in unstable air.

Contact Chamber. A closed pipe chamber constructed to permit the introduction of chemicals to treat fluids for specified periods of time.

Contact Condenser. A condenser in which coolant, vapors, and condensate are mixed.

Contact Pesticide. A chemical that kills pests on contact with the body, rather than by ingestion (stomach poison).

Contaminant. A harmful or undesirable constituent. In compressed respiratory air, any substance that might constitute a health hazard.

Contamination. A general term signifying the introduction into water of microorganisms, chemicals, wastes, or sewage which renders the water unfit for its intended use.

Continuous-Feed Incinerator. An incinerator into which refuse is charged in a nearly continuous manner in order to maintain a steady rate of burning.

Contrails. Long narrow clouds caused by the disturbance of the atmosphere during passage of high-flying jets. Proliferation of contrails may cause changes in the weather.

Control Agency (Air Pollution). State or local agencies with designated authority as official air pollution control bodies for those areas.

Control Program. Those activities and functions within a broader program of air resource management that collectively are directed toward the reduction of excessive emissions of pollutants; the regulatory aspects of an air resource management program.

Control Strategy. A combination of measures designed to achieve the aggregate reduction of emissions necessary for attainment and maintenance of a national ambient air quality standard.

Control Systems. Operating procedures or devices specifically designated to achieve the aggregate reduction of emissions for attainment of air standards by reducing the amount of pollutants emitted to the atmosphere.

Convection. Mixing motions in a fluid arising from the conversion of potential energy of hydrostatic instability into kinetic energy. It is more precise to term this motion "free convection" to distinguish it from "forced convection," which arises from external forces.

Converter. A furnace in which air is blown through a bath of molten metal or matter, oxidizing the impurities and maintaining the temperature through the heat produced by the oxidation reaction. In nitric acid production, the chamber in which ammonia is converted to nitric oxide and water by reacting it with air over a platinum-rhodium catalyst.

Converting. The process of removing impurities from molten metal or metallic compounds by blowing air through the liquid. The impurities are changed either to gaseous compounds, which are removed by volatilization, or to liquids or solids which are removed as slags.

Coolant. A substance, usually liquid or gas, used for cooling any part of a reactor in which heat is generated, including the core, the reflector, shield and other elements that may be heated by absorption of radiation.

Cooling Degree Days (CDD). A measure of the warmth of a region. To compute the CDD, average the high and low temperature of each day and add the resultant number of such days per year which are more than 65°F.

Cooling Tower. A device to remove excess heat from water used in industrial operations, notably in electric power generation.

Copper (Cu). A reddish colored metallic element with an atomic weight of 63.57 and atomic number of 29. Used extensively in its elemental form as well as in compounds in thousands of uses.

Core. The heart of a nuclear reactor where energy is released; also, the central part of a sand mold as used in foundries. The device placed in a mold to make a cavity in a casting.

Core Binder. Organic material added to foundry sand to aid in formation of a strong core for casting. Flour, linseed, oil, starch, and resins are among materials used.

Coriolis Force. A deflecting force caused by the earth's rotation on its axis.

Corrasion. The mechanical process involved in the wearing away of rock and soil by the scouring action of rock fragments activated by wind or water.

Corrosion. The gradual deterioration or destruction of a substance or material by chemical action or electrochemical process.

CPK. Creatine phosphokinase.

CPM. Critical Path Method – used in planning; see PERT.

CPU. Central Processing Unit, the main frame or core of a computer.

Crenothrix. A genus of bacteria characterized by unbranched attached filaments having a gelatinous sheath in which iron is deposited. They precipitate metallic deposits in pipes which may discolor the water and after their death cause disagreeable taste and odor in the water.

Countercurrent Ion Exchange. Refinement of the ion exchange process in which a countercurrent flow of raw water permits the most freshly regenerated resin to function as a "polishing" layer to remove the last traces of impurity from treated water.

Cover Material. Soil that is used to cover compacted solid waste in a sanitary landfill.

CPSC. U.S. Consumer Product Safety Commission.

CR. Congressional Record.

Cr. See Chromium.

Cracking. Chemical reaction by which large oil molecules are decomposed into smaller, lower-boiling molecules. At the same time, certain of these molecules, which are reactive, combine with one another to give even larger molecules than those in the original stock. The more stable molecules leave the system as cracked gasoline, but the reactive ones polymerize, forming tar and even coke. Cracking may be in either the liquid or vapor phase. When a catalyst is used to bring about the desired chemical reaction, this is called "catalytic cracking;" otherwise, it is assumed to be "thermal cracking" (see catalytic cracking).

Cracklings. The crisp residue left after the fat has been separated from the fibrous tissue in rendering lard or frying or roasting the skin of pork, turkey, duck, or goose.

Criteria Pollutants. Selected and specified polluting materials for which limiting standards are established.

Critical Wind Speed. The wind speed at the height of an elevated plume for which the maximum ground concentration is highest in neutral conditions.

Cross Connection. In plumbing, a connection through which a supply of potable water could be contaminated, polluted or infected.

Croy. A protective barrier built out into a stream to prevent erosion of a bank at a particular point.

CRSTER. A mathematical model whose computer output portrays the approximate short term air quality conditions in an area under varying conditions.

Crusher. A machine for crushing rock or other materials. Among the various types of crushers are the ball mill, gyratory crusher, Hadsel mill, hammer mill, jaw crusher, red mill, rolls, and stamp mill.

Cryolite. Sodium aluminum fluoride (Na_3AlF_6) used as an electrolyte in smelting of alumina to provide aluminum.

Crystallizer. A device using heat transfer with boiling to remove water, giving higher concentration of dissolved solids, usually less than 50%. Product may be distilled water and/or concentrated effluent.

CS. Cirrostratus cloud formation.

Cs. See Cesium.

C-Scale. Noise frequency scale.

CSRS. Cooperative State Research Service (USDA).

CSSO. Criteria and Special Studies Office (EPA).

CST. Crude sulfate turpentine.

CTG. Control Technology Guideline, EPA.

Cu. See Copper.

cu. Cumulus cloud formation.

cu. ft. Cubic feet.

Culm. The fine refuse from anthracite coal production.

Cultural Eutrophication. Acceleration by man of the natural aging process of bodies of water.

Culvert. A conduit for the free passage of surface drainage water under a highway, railroad, canal, or other embankment.

Cumulose. An organic matter soil deposit with some small amounts of rock waste included.

Cunnette. A longitudinal trough or channel built as an integral part of a large flat-bottomed conduit for concentrating the low flows to develop self-cleansing velocities.

Cupola. A vertical shaft furnace used for melting metals, especially grey iron, by having the charge come in contact with the hot fuel, usually metallurgical coke. Metal, coke, and flux are charged from the top of the furnace onto a bed of hot coke through which air is blown.

Curie. A measure of radioactivity.

Curium (Cm). An element. Atomic weight (most stable isotope) 247; atomic number 96.

Curtain-Wall. A partition wall between chambers in an incinerator under which combustion gases pass.

Cutie-Pie. A portable instrument equipped with a direct reading meter used to determine the level of radiation in an area.

cv. Closing volume.

CXT. Exposure concentration in ppm multiplied by time of exposure in hours.

Cycle of Fluctuation. The total time occupied by a period of rise and a succeeding period of decline of a water table. The most common are daily, annual and secular.

Cyclone. A structure without moving parts in which the velocity of an inlet gas stream is transformed into a confined vortex from which centrifugal forces tend to drive the suspended particles to the wall of the cyclone body. The particles then slide down the cyclone wall and are collected at the bottom.

Cyclone Collector. A device used to collect large-size particulates from polluted air by centrifugal force.

Cyclone Scrubbers. Devices ranging from simple dry cyclones with spray nozzles to multistage devices. All feature a tangential inlet to a cylindrical body.

Cyclonic Spray Tower. Liquid scrubbing apparatus where sprays are introduced countercurrent to gases for removal of contaminants.

Cyclotron. One kind of accelerator.

CZMA. Coastal Zone Management Act.

D

D. Weather abbreviation for "dust."

Damper. A manually or automatically controlled valve or plate in a breeching, duct, or stack, that is used to regulate a draft or the rate of flow or air or other gases.

Daughter. In nuclear fission, the product nucleus or atom that results from the decay of the parent.

dB. See decibel.

DBCP. Dibromochloropropane; a chemical compound used in pesticides; alleged to cause sterility in humans.

dcf. Dry cubic feet.

dcm. Dry cubic meter.

DCO. Delayed Compliance Order (EPA).

D=ct. Dosage equals concentration multiplied by time.

DDT. The first of the modern chlorinated hydrocarbon insecticides whose chemical name is 1,1,1-tricholoro-2,2-bs (p-chloriphenyl)-ethane. It has a half-life of 15 years, and its residues can become concentrated in the fatty tissues of certain organisms, especially fish. Because of its persistence in the environment and its ability to accumulate and magnify in the food chain, EPA has banned the registration and interstate sale of. DDT for nearly all uses in the United States effective December 31, 1972.

Decibel (dB). A unit of sound measurement.

Decomposition. Reduction of the net energy level and change in chemical composition of organic matter because of the actions of aerobic or anaerobic microorganisms.

Deconcentrator. An apparatus for removing suspended solids from boiler feed water.

Deferrization. The removal of soluble compounds of iron in water treatment.

Deflation. The removal of loose material by the wind, leaving the rocks bare to the continuous attack of weather.

Degasification. The removal of oxygen from water to lessen its corroding power; it may be accomplished by mechanical means, chemical methods, or a combination of both.

Degradation. The geologic process by means of which various parts of the surface of the earth are worn down and carried away and their general level lowered by the action of wind and weather; also the lowering of air quality in an area.

Degreasing. The process of removing grease and oils from sewage, waste, sludge or garbage.

Degree Days. The number of days a region is hotter or colder than 65°F – usually computed for annual variations.

Dehydration. Removal of water and water vapor by any means from a gas, liquid or solid.

Dehydrogenation. The removal of hydrogen from a chemical compound; for example, the removal of two hydrogen atoms from butane to make butylene, and the further removal of hydrogen to make butadiene.

DEIS. Draft Environmental Impact Statement.

Del Sub. Deleterious substances.

Delf. A drain behind a sea embankment on the land side.

Delta. The alluvial deposit where a river or stream drops its debris on entering a body of quieter water. A river mouth area where many branches form islands of alluvial deposits.

Demanganization. In water treatment, the removal of compounds of manganese from water.

Demineralization. Removal from water of mineral contaminants. Methods include ion exchange, flash distillation, electrodialysis, reverse osmosis, and precipitation by settling.

Demister (Collector). A mechanical device used to eliminate finely divided liquid particles from process streams by impaction and agglomeration. Apparatus made of wire mesh or glass fiber and used to eliminate acid mist as in the manufacture of sulfuric acid.

Dendritic. The form of the drainage pattern of a stream and its tributaries when it follows a tree-like shape with main trunk, branches and twigs as sub-tributaries.

Denitrification. The reduction of nitrates in solution by bio-chemical action.

Densifier. A component of a waste recycle plant which moistens and presses shredded wastes into "energy cubes."

Denudation. The erosion by rain, frost, wind, running water, and other agencies of the solid matter of the earth so that strata formerly covered are exposed. Also, the leveling of vegetation including forests by similar action or by fire.

Deoxygenation. The depletion of the dissolved oxygen in a liquid under natural conditions of the biochemical oxidation of organic matter present.

Depletion. The continued withdrawal of groundwater or water from an acquifer or surface storage area at a rate greater than replenishment.

Dermal Toxicity. The ability of a pesticide chemical to poison an animal or human by skin absorption.

Desalinization. Salt removal from sea or brackish water.

Desiccant. A chemical agent that may be used to remove moisture from plants or insects causing them to wither and die.

Destructive Distillation. A process of distillation in which an organic compound or mixture is heated to a temperature high enough to cause decomposition. The heating of organic matter when air is not present, resulting in the evolution of volatile matter and leaving char consisting of fixed carbon and ash.

Desulfurization. In coal processing, the removal of sulfur from the coal, often by mechanical cleaning processes. In petroleum refining, removing sulfur or sulfur compounds from a charge stock (oil that is to be treated in a particular unit).

Detention Time. The theoretical time required to displace the contents of a tank or unit at a given rate of discharge. Volume divided by rate of discharge.

Detergent. Synthetic washing agent that, like soap, lowers the surface tension of water, emulsifies oils and holds dirt in suspension. Environmentalists have criticized detergents because most contain large amounts of phosphorus-containing compounds that contribute to the eutrophication of waterways.

Detritus. A large grit detention chamber usually with provision for removing the sediment without interrupting the flow of liquid.

Deuterium. See heavy hydrogen.

Dewpoint. The temperature, at atmospheric pressure, at which air and gases are saturated with moisture, and the water vapor they hold begins to condense or deposit as a liquid.

Diapers. Slang term for drainage silt curtain.

Diatomaceous Earth (Diatomite). A fine siliceous material resembling chalk used in waste water treatment plants to filter sewage effluent to remove solids. May also be used as inactive ingredients in pesticide formulations applied as dust or powder.

Dibromochloropropane. See DBCP.

Diesel Engine. A type of internal-combustion engine that uses a fuel injector and produces combustion temperatures by compression.

Diffused Air. A type of sewage aeration. Air is pumped into the sewage through a perforated pipe.

Diffuse Radiation. The solar radiation which strikes an object after the radiation has been scattered and diverted by particles in the atmosphere.

Diffusion. The spreading or scattering of a gaseous or liquid material. Eddy diffusion: Diffusion caused by turbulent activity in a fluid system. Molecular diffusion: A process of spontaneous intermixing of different substances, attributed to molecular motion and tending to produce uniformity of concentration.

Digester. In a waste water treatment plant, a closed tank that decreases the volume of solids and stabilizes raw sludge by bacterial action.

Digester Coils. A system of pipes for hot water or steam installed in sludge-digestion tank for the purpose of heating the sludge.

Digestion. The biochemical decomposition of organic matter. Digestion of sewage sludge takes place in tanks where the sludge decomposes, resulting in partial gasification, liquefaction and mineralization of pollutants.

Dilution Ratio. The ratio of the volume of water of a stream to the volume of incoming waste. The capacity of a stream to assimilate waste is partially dependent upon the dilution ratio.

Dioxirane. Three ring molecule made up of carbon, oxygen and hydrogen atoms formed as an intermediate in the reaction between ozone and ethylene.

Direct Conversion. Conversion of sunlight directly into electric power.

Direct Radiation. That radiation which passes through the atmosphere without deflection or scattering due to molecules in the air.

Direct-fired Boiler. Commonly used to denote a boiler and furnace fired by pulverized coal.

Disinfection. Effective killing by chemical or physical processes of all organisms capable of causing infectious disease. Chlorination is the disinfection method commonly employed in sewage treatment processes.

Dispersant. A chemical agent used to break up concentrations of organic material. In cleaning oil spills, dispersants are used to disperse oil from the water surface.

Dispersion. The dilution or removal of a substance by diffusion turbulence, and mixing. Involves two substances, the first of which is uniformly distributed in a finely divided state through the second (the dispersion medium).

Dissolved Oxygen (DO). The oxygen dissolved in water or sewage. Adequately dissolved oxygen is necessary for the life of fish and other aquatic organisms and for the prevention of offensive odors. Low dissolved oxygen concentrations generally are due to discharge of excessive organic solids having high BOD, the result of inadequate waste treatment.

Dissolved Solids. The total amount of dissolved material, organic and inorganic, contained in water or wastes. Excessive dissolved solids make water unpalatable for drinking and unsuitable for industrial uses.

Distillate. The product of distillation obtained by condensing the vapors from a still.

Distillate Fuels. Liquid fuels distilled usually from crude petroleum, except residuals such as No. 5 and No. 6 fuel oil.

Distillate Oils. The lighter oils produced by distilling crude oil.

Distillation. The process of heating a substance to the temperature at which it is converted to a vapor, then cooling the vapor, and thus restoring it to the liquid state.

Diurnal. Twice daily; cyclic events occurring twice in a 24 hour period such as the tides.

Divisor. A device used in soil conservation studies which controls the continuous separation and diversion of a representative fraction of the runoff from experimental areas.

DMNA. Dimethylnitrosamine.

DNA. Deoxyribonucleic acid.

D.O. See dissolved oxygen.

Doctor Treatment. Treatment of gasoline with sodium-plumbite solution and sulfur to improve its odor.

DOD. Department of Defense.

DOE. Department of Energy.

DOI. Department of the Interior.

Dopes for Gasoline. Materials added in small amounts to gasoline to increase the octane number and thus help to prevent knocking.

Dose. In radiology, the quantity of energy or radiation absorbed.

Dosimeter (Dosemeter). An instrument used to measure the amount of radiation a person has received.

Dosing Tank. A tank into which raw or partly treated sewage is introduced and held until the desired quantity has been accumulated after which it is discharged at a specified rate for proper treatment.

DOT. Department of Transportation.

Double Decomposition. A chemical reaction between two compounds in which part of the first compound becomes united with the remainder of the second, as: AB + CD = AD + BC.

Downpass. A chamber or gas passage placed between two combustion chambers to carry the products of combustion downward.

Downwash. The downward motion of part or all of a plume due to the lower pressure in the wake of the stack or building or due to a downward step of the terrain.

Draft. The difference between the pressure in an incinerator, or any component part and that in the atmosphere; it causes air or the products to flow from the incinerator to the atmosphere. Also, the depth a vessel sinks in water when floating free measured from the waterline to the keel or bottom. Also, the preliminary working copy of a written statement.

Draft Controller. An automatic device that maintains a uniform furnace draft by regulating a damper.

Draw. A natural depression or swale.

Dredging. A method for deepening streams, swamps or coastal waters by scraping and removing solids from the bottom. The resulting mud is usually deposited in marshes in a process called filling. Dredging and filling can disturb natural ecological cycles. For example, dredging can destroy oyster beds and other aquatic life; filling can destroy the feeding and breeding grounds for many fish species.

DRI. Development of Regional Impact.

Drift. The amount or rate of a water current expressed in knots (nautical miles per hour) or in miles (statute) per hour – also, water lost to the atmosphere through the louvered sides of a cooling tower.

Drillers Mud. Slang term for drilling fluid in well construction work.

Drilling Fluid. A viscous liquid made up of water, clay and/or bentonite or other additives, used to cool the drill bit and flush the chips from the hole in well drilling.

Drill Pipe. Hollow shaft used to drill wells. It extends from a swivel attached to a winch through a rotary drive table and terminates in a drill bit. The swivel has a hose attached which supplies drilling fluid to the ports in the drill bit.

Drill String. Slang term for column of drill pipe in well construction work.

Drop Arch. A refractory construction or baffle which serves to deflect gases in a downward direction.

Dross. Impurity formed in melted metal. A zinc-and-iron alloy forming in a bath of molten zinc, in galvanizing iron. The scum that forms on the surface of molten metals usually due to oxidation, but occasionally due to the rising of impurities to the surface.

Drum, Flash (or Flash Tower). A drum or tower into which the heated outlet products of a preheater or exchanger system are conducted, often with some release in pressure. The purpose of the drum is to allow vaporization and separation of the volatile portions for fractionation elsewhere.

Dry Bottom Furnace. A furnace designed to burn pulverized coal at temperatures low enough to prevent the ash from fusing or slagging.

Dry Limestone Process. A method of controlling air pollution caused by sulfur oxides. The polluted gases are exposed to limestone which combines with oxides of sulfur to form manageable residues.

DRZL. Weather abbreviation for drizzle.

dscf. Dry cubic feet at standard conditions.

DSSE. Division of Stationary Source Enforcement (EPA HQ element).

Duff. That layer of undecomposed organic matter lying on the surface of the ground as in a forest.

Dump. A land site where solid waste is disposed of in a manner that does not protect the environment.

Dust. Generally particles from 1 to 100 micro-meters in size that become airborne by natural or mechanical means. These particles do not diffuse but will settle under the influence of gravity (see also particle).

Dust Collecting Fan. A centrifugal fan which concentrates dust and skims it into a cyclone or hopper.

Dustfall Jar. An open-mouthed container used to collect large particles that fall out of the air. The particles are measured and analyzed.

Dustless Loading. The amount of dust in a gas, usually expressed in grains per cubic foot or in pounds per thousand pounds of gas (see also grain loading).

Dy. See Dysprosium.

Dysprosium (Dy). A rare earth element whose atomic weight is 162.46 and atomic number is 66. It forms highly colored salts.

Dystrophic Lakes. Lakes between eutrophic and swamp stages of aging. Such lakes are shallow and have high humus content, high organic matter content, low nutrient availability and high BOD.

E

EAS. Environmental Assessment Statement.

Ebb Tide. The seaward flow from estuarine areas of waters caused by the tidal infuence of the moon.

EBDC. Ethylene bisdithio carbamates – a pesticide alleged to be carcinogenic.

ECHO. Each Community Helps Others, a noise abatement program sponsored by EPA.

Ecological Impact. The total effect of an environmental change, either natural or man-made, on the ecology of the area.

Ecology. The interrelationships of living things to one another and to their environment or the study of such interrelationships.

Economic Poisons. Those chemicals used to control insects, rodents, plant diseases, weeds and other pests, and also to defoliate economic crops such as cotton.

Economizer. A heat recovery device designed to transfer heat from the products of combustion to a fluid, usually feedwater for a steam boiler. The water flows through a bank of tubes placed across the flue gases and is heated by these gases prior to entering the boiler.

Ecosphere. See biosphere.

Ecosystem. The interacting system of a biological community and its non-living environment.

ECSL. Enforcement Compliance Schedule Letter (USEPA).

EDB. Ethylene dibromide.

Eddy. Circular movements occurring in air or water flow; as the eddies of air current caused by structures or hills, or the eddies caused in streams by obstructions or irregularities in the banks or bottoms.

EDF. Environmental Defense Fund.

Eductor. A device for mixing air with water.

Effective Half-Life. See biological half-life.

Effective Stack Height. Variously defined. The three most common definitions are (1) the height at which a plume levels off, which has been observed only in stable conditions; (2) the height of a plume above the point of maximum ground concentration; (3) the virtual height of plume origin based on the diffusion pattern at large distances downwind of the stack. Definition 1 is the easiest to apply in stable conditions; definition 2 is the most practical in neutral and unstable conditions; definition 3 is comprehensive but difficult to apply.

Efficiency. The ratio of output to input. The efficiency of a stream generating unit is the ratio of the heat absorbed by the water or steam to the heat in the fuel fired, expressed in per cent.

Efflorescence. The release of moisture upon exposure to the atmosphere.

Effluent. A discharge of pollutants into the environment, partially or completely treated or in its natural state. Generally used in regard to discharges into water.

Effluent Seepage. Diffuse discharge onto or into the ground of liquids that have percolated through solid waste or another medium; they contain dissolved or suspended materials.

Effluent Water Separator. A container designed to separate volatile organic compounds from waste water prior to discharge or reuse.

Efflux Velocity. The mean speed of exiting stack gases.

EI. Emission Inventory, EPA.

EIA. Electronics Industries Association.

Einsteinium (Es). Man-made radio-active element. Atomic weight (most stable known isotope) 254; Atomic number 99.

EIS. See Emissions Inventory System; Also Environmental Impact Statement.

Ejector. A device of moving a fluid or solid by entraining it in a high velocity steam, air or water jet.

Ekatantalum. See Protoactinium.

EKG. Electrocardiogram.

EKMA. Empiric Kinetic Modeling Approach for HC, EPA.

Electrical Energy. The energy of an electric current or electromagnetic waves.

Electrode. A conductor used to establish electrical contact with a non-metallic part of a circuit, such as an electrolyte. There are two electrodes in such a circuit, the anode, with a positive charge, and the cathode, with a negative one. See electrolysis.

Electrodialysis. A process that uses electrical current and an arrangement of permeable membranes to separate soluble minerals from water. Often used to desalinize salt or brackish water.

Electrolysis. A form of corrosion occurring in metallic pipes or structures by stray electric currents or galvanic action electron flow of dissimilar metals; also the production of chemical changes by means of an electric current passing through an electrolyte.

Electrolyte. A non-metallic substance which will conduct an electric current by the movement of ions when dissolved in certain solvents or when fused by heat; common salt is an example.

Electrolytic Cell. A device for carrying out an electrochemical reaction.

Electromagnetic Waves. Waves of radiant energy, commonly classified, according to wave frequency and length, as Hertzian (radio), infrared, visible (light), ultraviolet, X-ray, and gamma ray; and, by extension, particle emissions, such as alpha and beta radiation, or rays of mixed or unknown type, as cosmic radiation.

Electron. See atom.

Electrostatic Precipitator. An air pollution control device that removes particulate matter by imparting an electrical charge to particles in a gas stream for mechanical collection on an electrode.

Elutriation. Separation of solid waste into heavy and light fractions by washing.

Elutriator. A vertical tube through which a gas or fluid passes upward at a specific velocity while a solid mixture whose separation is desired is fed into the top of the column. The large particles which settle at a velocity higher than that of the rising fluid are collected at the bottom of the column, and the smaller particles are carried out of the top of the column with the fluid.

Eluviation. A process of removal of organic material in solution or in suspension from the soil by percolating waters.

EMARS. Energy Materials Activity Recommendation System (DOI).

Embayment. A deep depression in a shore line forming a large open bay.

Emergency Episode. See air pollution episode.

Emission. The total amount of solid, liquid, or gaseous pollutant emitted into the atmosphere from a given source in a given time, and indicated in grams per cubic meter of gas, pounds per hour, or other quantitative measurement.

Emission Factor. The average amount of a pollutant emitted from each type of polluting source in relation to a specific amount of material processed. For example, an emission factor for a blast furnace (used to make iron) would be a number of pounds of particulates per ton of raw materials.

Emission Inventory. A list of air pollutants emitted into a community's atmosphere, in amounts (usually tons) per day, by type of source. The emission inventory is basic to the establishment of emission standards.

Emission Point/Discharge Point. The point of temporary, intermittent, or continuous release of foreign matter to the air, or matter that is common to the air but in an amount, form, or location such as to give the common matter special significance.

Emission Standard. The maximum amount of a pollutant legally permitted to be discharged from a single source, either mobile or stationary.

Emissivity. The proportion of the incident energy that is reradiated from a surface. (Note: not reflected).

EMSL. Environmental Monitoring and Support Laboratory (EPA).

Emulsoid. A colloid which is readily dispersed in a suitable medium and may be redispersed after coagulation.

Emphysema. See pulmonary emphysema.

Energy. The capacity to do work. It may take a number of forms, among them mechanical, chemical, and radiant, and can be transformed from one form to another, but cannot be created or destroyed.

Endothermic Reaction. A reaction which requires the addition of heat for its continuation.

Engine. A machine for producing mechanical energy.

Enrichment. The addition of nitrogen, phosphorus and carbon compounds or other nutrients into a lake or other waterway that greatly increases the growth potential for algae and other aquatic plants. Most frequently, enrichment results from the inflow of sewage effluent or from agricultural runoff.

Entrainment. The dilution of plume properties due to mixing with the ambient fluid.

Environment. The sum of all external conditions and influences affecting the life, development and, ultimately, the survival of an organism.

Environmental Heat. Heat from the sun stored in substances such as rocks, water or air.

Environmental Impact Statement. A document prepared by a Federal agency on the environmental impact of its proposals for legislation and other major actions significantly affecting the quality of the human environment. Environmental Impact Statements are used as tools for decision making and are required by the National Environmental Policy Act.

EOOS. Equipment out of Service.

EPA. United States Environmental Protection Agency.

Epidemiology. The study of diseases as they affect populations.

Epilimnion. Upper layer of water in a lake, being warmer and less dense than lower waters.

EPRI. Electric Power Research Institute.

eq. Equivalent.

Equivalent Method. Any method of sampling and analyzing for air pollutant which has been demonstrated to the administrators (EPA) satisfaction to have a consistent and quantitatively known relationship to the reference method under specified conditions.

Er. See Erbium.

Erbium (Er). A rare earth element whose atomic weight is 167.64 and atomic number is 68. It forms highly colored salts.

ERC. Environmental Reporter Cases.

Erg. Unit of energy measurement.

EROS. Earth Resources Observations System (National Space Technology Laboratories USGS).

Erosion. The wearing away of the land surface by wind or water. Erosion occurs naturally from weather or run-off but is often intensified by man's land-clearing practices.

ERS. Economic Research Service (USDA).

ER's. Environmental reports.

ERTS. Earth Resources Technology Satellite.

Es. Einsteinium.

ESECA. Energy Supply and Environmental Coordination Act.

ESRL. Environmental Services Research Laboratory (EPA).

Estuaries. Areas where the fresh water meets salt water. For example, bays, mouths of rivers, salt marshes and lagoons. Estuaries are delicate eco-systems; they serve as nurseries, spawning and feeding grounds for a large group of marine life and provide shelter and food for birds and wildlife.

Estuarine. See estuaries.

Eu. See Europium.

Europium (Eu). A rare earth element whose atomic weight is 152.0 and atomic number is 63. It forms highly colored salts.

Eutrophication. The normally slow aging process by which a lake evolves into a bog or marsh and ultimately assumes a completely terrestrial state and disappears. during eutrophication the lake becomes so rich in nutritive compounds, especially nitrogen and phosphorus, that algae and other microscopic plant life become superabundant, thereby "choking" the lake, and causing it eventually to dry up. Eutrophication may be accelerated by many human activities.

Eutrophic Lakes. Shallow lakes, weed-choked at the edges and very rich in nutrients. The water is characterized by large amounts of algae, low water transparency, low dissolved oxygen and high BOD.

Evaporation. The process by which a liquid passes from a liquid state to a vaporous state below the boiling point of the liquid.

Evaporation Ponds. Shallow, artificial ponds where sewage sludge is pumped, permitted to dry and either removed or buried by more sludge.

Evaporative Cooler. A bare evaporator coil through which cool air passes and over which water is sprayed to accomplish heat transfer in the water through evaporation.

Evaporator. Usually a vessel which receives the hot discharge from a heating coil and, by a reduction in pressure, flashes off overhead the light products and allows the heavy residue to collect in the bottom (see flash tower).

Evapo-transpiration. Water withdrawn from soil by evaporation and plant transpiration.

Evase Stack. An expanding connection on the outlet of a fan or in an air flow passage; its purpose is to convert kinetic energy into static pressure.

EW. Weather abbreviation for "sleet showers."

Excess Air. Air supplies for combustion in excess of that theoretically required for complete combustion, usually expressed as a percentage of theoretical air, such as "130 per cent excess air."

Excess Emissions. Emission of pollutants in excess of those allowed by process weight limits, specific emission limiting regulation, or permit conditions.

Existing Facility. See existing source.

Existing Source. A source which is in existence, operating or under construction or has received a permit to construct prior to a specified target date.

Exothermic Reaction. A reaction which produces heat.

F

F. Weather abbreviation for "fog."

°F. Degrees Fahrenheit.

Fabric Filters. A device for removing dust and particulate matter from emissions much like a home vacuum cleaner bag. The most common use of fabric filters is the baghouse.

Fanning. The term applied to the effect of an inversion on a plume rise causing it to diffuse laterally rather than vertically.

FCC. Federal Communications Commission.

FDA. Food and Drug Administration.

Fe. See Iron.

FEA. Federal Energy Administration (obs.).

Fecal Coliform Bacteria. A group of organisms common to the intestinal tracts of man and of animals. The presence of fecal coliform bacteria in water is an indicator of pollution and of potentially dangerous bacterial contamination.

Federal Reference Method (FRM). Any method of sampling and analyzing for an air pollutant, as established by the U.S. EPA under Federal law.

Feedlot. A relatively small, confined land area for raising cattle. Although an economical method of fatteningbeef, feedlots concentrate a large amount of animal wastes in a small area. This excrement cannot be handled by the soil as it could be if the cattle were scattered on open range. In addition, runoff from feedlots contributes excessive quantities of nitrogen, phosphorus and potassium to nearby waterways, thus contributing to eutrophication.

Feedstock. Starting material used in a process. This may be raw material or an intermediate product that will undergo additional processing.

Feed Water. Distilled water pumped into boilers to take place of that vaporized into steam.

FEF. Forced expiratory flow.

Fen. A low-lying land area partly covered by water.

FERC. Federal Energy Regulatory Commission.

Fermium (Fm). Man-man radio-active element. Atomic weight (most stable known isotope) 257; atomic number 100. Produced by neutron bombardment of plutonium.

FET. First edge time.

Fetch. An open stretch of water in a stream, river, or harbor fairway.

FEV. Forced expiratory volume.

FGD. Flue gas desulfurization.

FHA. Federal Housing Administration.

FIFRA. Federal Insecticide, Fungicide, and Rodenticide Act.

Filling. The process of depositing dirt and mud in marshy areas to create more land for real estate development. Filling can disturb natural ecological cycles. See dredging.

Film Badge. A piece of masked photographic film worn like a badge by nuclear workers to monitor an exposure to radiation. Nuclear radiation darkens the film.

Filter Collector. A mechanical filtration system for removing particulate matter from a gas stream, for measurement, analysis, or control. Also called bag collector. Filters are designed in a variety of sizes and materials for specific purposes.

Filtration. In waste water treatment, the mechanical process that removes particulate matter by separating water from solid material usually by passing it through sand.

Filtration Rate. The rate at which liquid is fed through a filter, in gallons/minute/sq. ft. of filter-media surface, qualified as standard, low and high.

Final Rise. The total plume rise after leveling off, if this occurs especially as opposed to the term "transitional rise."

Fines. Fine particulates; aerosols.

Fish Ladder. See fishway.

Fishway. A device constructed in connection with a dam to allow for migration of fish between two levels.

Fission. The splitting of atomic nuclei into smaller nuclei, accompanied by the release of great quantities of energy.

Fixed Carbon. The ash-free carbonaceous material that remains after volatile matter is driven off during the analysis of a dry solid waste sample.

Flare. A fire produced by a pilot flame at a stack outlet, used in many heat treating operations and petroleum industry processes to burn combustible waste gases.

Flareback. A burst of flame from a furnace in a direction opposed to the normal gas flow; it usually occurs when accumulated combustible gases ignite.

Fleam. An artificial watercourse.

FLM. Federal Land Manager, EPA.

FLMPA. Federal Land Management and Policy Act.

Float Control. A float installed in a tank or body of liquid to control the flow by pumps or mechanical actuators.

Floating Roof. A special tank roof which floats upon the oil in a storage tank.

Floc. A clump of solids formed in sewage by biological or chemical action.

Flocculation. In waste water treatment, the process of separating suspended solids by chemical creation of clumps or flocs.

Flood Prone Area. A low area near a body of water which has a history (usually taken over past 100 years) of flooding. Insurance rates in flood prone areas are necessarily higher. The USGS prepares charts showing flood prone areas.

Flood Tide. The shoreward flow into estuarine areas of ocean waters caused by the tidal influence of the moon.

Flotsam. Debris and loose material floating on the surface of waters.

Flowmeter. In waste water treatment, a meter that indicates the rate at which waste water flows through the plant.

Flue. Any passage designed to carry combustion gases and entrained particulates.

Flue Dust. Solid particles, smaller than 100 microns, carried in the products of combustion.

Flue Gas. A mixture of gases resulting from combustion and emerging from a chimney. Flue gas includes nitrogen oxides, carbon oxides, water vapor and often sulfur oxides or particulates.

Flue Gas Scrubber or Washer. Equipment that removes fly ash and other objectionable materials from flue gas by the use of sprays, wet baffles, or other means that require water as the primary separation mechanism.

Fluid Bed Processor. A device in which solid particles are suspended in a gaseous media, and exhibit properties of a fluid. Used for drying, cooling, dehydration, purification, reaction, and particle growth.

Fluidized Bed Technique. A combustion process in which heat is transferred from finely divided particles, such as sand, to combustible materials in a combustion chamber. The materials are supported and fluidized by a column of moving air.

Fluidized Roasting. Oxidation of finely ground pyritic minerals by means of upward currents of air, blown through a reaction vessel (fluid bed roaster) with sufficient force to cause the bed of material to expand (boil). Reaction between mineral and air is maintained at a desired exothermic level by control of oxygen entry, by admission of cooling water, or by addition of fuel.

Flume. A channel, either natural or man-made, which carries water.

Fluorides. Gaseous, solid or dissolved compounds containing fluorine, emitted into the air or water from a number of industrial processes. Fluorides in the air are a cause of vegetation damage and, indirectly, of livestock damage.

Fluorine (F). A halogenic element which vaporizes as a pale yellow gas and whose atomic weight is 19.00 and atomic number is 9. It is used in etching and in dental work. The element and its principal acid HF are dangerous poisons.

Fluorosis. A chronic poisoning resulting from the presence of 0.9 milligrams or more per liter of fluorine in drinking water. Teeth become brittle and opaque white with a mottled enamel.

Fluorospar. A natural calcium fluoride (CaF_2) used as a flux in open hearth steel furnaces and in gold, silver, copper, and lead smelting.

Fluxing. Dissolving or melting of a substance by chemical action.

Fly Ash. All solids, including ash, charred paper, cinders, dust, soot or other partially incinerated matter, that are carried in a gas stream.

Fm. Fermium.

FMVCP. Federal Motor Vehicle Control Program.

Focusing Collector. Focusing collectors utilize optical systems and/or concentrating reflector surfaces to increase the intensity of solar radiation on the energy-absorbing surface.

Fog. Liquid particles formed by condensation of vaporized liquids.

Fogging. The application of a pesticide by rapidly heating the liquid chemical, thus forming very fine droplets with the appearance of smoke. Fogging is often used to destroy mosquitoes and blackflies.

FOI. Freedom of Information.

FOIA. Freedom of Information Act.

Food-Grade Acid. Phosphoric acid that has been treated for removal of heavy metals and is suitable for use in food products.

Food Waste. Animal and vegetable waste resulting from the handling, storage, sale, preparation, cooking and serving of foods; commonly called garbage.

Forced Circulation. A system that uses a fan or pump to create circulation of a fluid.

Forced Draft. The positive pressure created by the action of a fan or blower, which supplies the primary or secondary combustion air in an incinerator, or ventilation system.

Formazin Turbidity Units (FTU). A measure of water turbidity and are equivalent to Jackson Turbidity Units (See JTU).

Fossil Fuels. Coal, oil and natural gas; so-called because they are derived from the remains of ancient plant and animal life.

Fossil Fuel Steam Generator. Furnaces and boilers which produce steam by combustion of oil, coal or gas of fossil origin.

Fouling. The impedence of the flow of gas or heat that results when material accumulates in gas passages or on heat-absorbing surfaces in an incinerator.

FPC. Federal Power Commission.

FPD. Flame photometric detectors.

FPS. Foot-pound-second system.

FR. Federal Register.

Fractional Distillation. The separation of the components of a liquid mixture by vaporizing and collecting the fractions which condense in different temperature ranges.

FRCU. Fracto cumulus cloud formation.

Freeboard. The vertical distance between the normal maximum level of the surface of the liquid in a conduit, reservoir, tank, canal and the top of the sides of such containment. It is provided so that wave action will not flood over the confining structure.

French Drain. Subterranean or surface excavated lines filled with gravel or other porous material emanating from a low saturated area to provide drainage where natural drainage is insufficient.

FRM. Federal Reference Method.

Froude Number. The ratio of pressure forces to buoyant forces. The efflux froude number of a stack may be defined as $W_0^2/[g(\text{delta } T/T)D]$.

FRRRPA. Forest and Rangelands Renewable Resources Planning Act.

FS. Forest Service (USDA); also a weather abbreviation for "Fractostratus cloud formation."

FSM. Fixed Station Monitoring.

ft. Feet.

FTU. See Formazin Turbidity Units.

Fuel. Any form of combustible matter – solid, liquid, vapor, or gas, excluding combustible refuse.

Fuel Bed. The layer of solid fuel or solid waste on a furnace grate or hearth.

Fuel-Burning Equipment. Any furnace, boiler, apparatus, stack, and all appurtenances thereto, used in the process of burning fuel for the primary purpose of producing heat or power by indirect heat transfer.

Fuel Cells. A device that uses a chemical reaction to produce electricity directly, such as the reaction of hydrogen and oxygen to produce electric power with water as a product.

Fugitive Dust. Dust created by a process, action, or motion of a property owner which leaves that property, by winds or by forced pressure. Solid airborne particulate matter other than from a flue or stack.

Fugitive Emissions. Particles which are generated by industrial or other activities and which escape to the atmosphere not through primary exhaust systems, but through openings such as windows, vents or doors, ill-fitting oven closures, or poorly maintained equipment.

Fugitive Particulate. Particulate matter which escapes and becomes airborne from unenclosed operations or which is emitted into the atmosphere without passing or being conducted through a flue pipe stack or other structure designed for the purpose of emitting air pollutants into the atmosphere.

Fume. Tiny solid particles commonly formed by the condensation of vapors of solid matter.

Fumigant. A pesticide that is burned or evaporated to form a gas or vapor that destroys pests. Fumigants are often used in buildings or greenhouses.

Fumigation. Fumigation is an atmospheric phenomenon in which pollution which has been retained by an inversion layer near its level of emission, is brought rapidly to ground level when the inversion breaks up. High concentrations of pollutant can thus be produced at ground level.

Fuming Nitric Acid. A mixture of 98 per cent nitric acid and an equilibrium mixture of nitrogen tetroxide (N_2O_4) and nitrogen dioxide (NO_2).

Fungi. Small, often microscopic plants without chlorophyll. Some fungi infect and cause disease in plants or animals; other fungi are useful in stabilizing sewage or in breaking down wastes for compost.

Fungicide. A pesticide chemical that kills fungi or prevents them from causing diseases, usually on plants of economic importance. See pesticide.

Furnace. The chambers of an incinerator where drying, ignition, and combustion occur.

Furnace Oil. A distillate fuel primarily intended for domestic heating use. No. 1 commercial standard grade is intended for "vaporizing" burners requiring a volatile fuel, whereas No. 2 and No. 3 commercial standard grades are less volatile, and are thus usable in the "atomizing" type of burners.

Fusion. The union of light nuclei to form heavier ones, accompanied by the release of enormous quantities of energy.

Fusion Point. The temperature at which a particular complex mixture of minerals can flow under the weight of its own mass. Because most refractory materials have no definite fusion points but soften gradually over a range of temperatures, the conditions of measurement have been standardized by ASTM (See Pyrometric Cone Equivalent).

FVC. Forced vital capacity.

FWPCA. Federal Water Pollution Control Act.

FWS. Fish and Wildlife Service (USDI).

FY. Fiscal Year.

G

g. Gram.

Gabion. A long wicker basket filled with earth or stones used in levee and dike work.

GAC. Granulated Activated Carbon.

Gage Pressure. The pressure above atmospheric pressure, expressed as pounds per square inch, gage (psig).

Gallery. An underground structure designed and installed for the purpose of collecting ground water.

Gadolinium (Gd). A rare earth element whose atomic weight is 157.3 and atomic number is 64. It forms salts which are highly colored.

gal(s). Gallon(s)

Game Fish. Those species of fish sought by sports fishermen; for example, salmon, trout, black bass, striped bass, etc. Game fish are usually more sensitive to environmental changes and water quality degradation than "rough" fish.

Gamma Ray. Waves of radiant nuclear energy. Gamma rays are the most penetrating of the three types of radiation and are best stopped by dense materials such as lead.

Garbage. See food waste.

Garbage Grinding. A method of grinding food waste by a household disposal, for example, and washing it into the sewer system. Ground garbage then must be disposed of as sewage sludge.

Gasification. The process of converting a solid or liquid fuel into a gaseous fuel. Also, the transformation of sewage solids into gas in the decomposition of sewage.

Gate. A movable watertight barrier for the control of a liquid in a waterway.

GC. Guanylate cyclase; also gas chromatograph.

GCT. Greenwich Civil Time.

Gd. See Gadolinium.

Geiger Counter. An electrical device that detects the presence of radioactivity.

Generation. The act or process of producing solid waste.

Generator. A device that converts mechanical energy into electrical energy.

GEOREF. World geographic reference system.

GEOS. NASA Geodetic Survey Satellite.

GEP. Good engineering practice.

g-eq. Grain equivalent.

Germicide. A chemical or agent that kills microorganisms such as bacteria and prevents them from causing disease. Such compounds must be registered as pesticides with EPA.

GF. Weather abbreviation for "ground fog."

Gl. See Beryllium.

Glucinium (Gl). See Beryllium.

GMP. Guanosine monophosphate.

GMT. Greenwich Mean Time.

Gob Piles. Large piles of low-combustible refuse from coal mine preparation plants. Fires may develop in these waste material piles by liberation of heat through slow oxidation, until ignition temperature is reached (see also culm).

gr. Grain.

Gradation. The geologic process of leveling a portion of the earth's surface; also the process of bringing the bed of a stream to a gradient at which the water flowing in it is just able to transport the material delivered to it.

Grade. Surface level; also the inclination of a stream bed, conduit, pipeline; also the elevation of the invert of the bottom of a pipeline, canal, culvert etc.

Gradient. The rate of change per unit of length – slope.

Grain. A unit of weight equivalent to 65 milligrams or 2/1,000 of an ounce.

Grain Loading. The rate of emission of particulate matter from a polluting source. Measurement is made in grains of particulate matter per cubic foot of gas emitted.

Grating. A screen consisting of parallel bars, two sets being transverse to each other in the same plane.

Gravitational Settling. Removal of material from the atmosphere due to the action of gravity.

Gravity Feed Systems. A system using gravity to move the medium. A solar system which utilizes the principle that heated water is less dense than cool water and therefore rises causing natural circulation.

Grease. In sewage, grease includes fats, waxes, free fatty acids, calcium and magnesium soaps, mineral oils and certain other non-fatty materials.

Grease Intercepter. A trap in a drain plumbing line designed to collect and retain grease and fatty substances.

Green Belts. Certain areas restricted from being used for buildings and houses; they often serve as separating buffers between pollution sources and concentrations of population.

Green Coke. Coke that has not been fully cooked. Green coke produces excessive emission when pushed from a coke oven.

Green Feed (Calcined Feed). Not fully processed or treated feed.

Greenhouse Effect. The heating effect of the atmosphere upon the earth. Light waves from the sun pass through the air and are absorbed by the earth. The earth then reradiates this energy as heat waves that are absorbed by the air, specifically by carbon dioxide. The air thus behaves like glass in a greenhouse, allowing the passage of light but not of heat. Thus many scientists theorize that an increase in the atmospheric concentration of CO_2 can eventually cause an increase in the earth's surface temperature.

Grit. Heavy, inorganic matter, such as sand or pebbles.

Groin. A wall, crib, row of piles, stone jetty or other barrier projecting out from a bank for the purpose of protecting the shoreline from erosion and to concentrate stream flow.

Ground Cover. Grasses or other plants grown to keep soil from being blown or washed away.

Ground Level Concentration. The mass per unit volume of solid, liquid, or gaseous material in micrograms per cubic meter of air, measured from 0 to 2 meters above the ground.

Groundwater. The supply of freshwater under the earth's surface in an aquifer or soil that forms the natural reservoir for man's use.

Groundwater Runoff. Groundwater that is discharged into a stream channel as spring or seepage water.

Grubbing. The operation of removing stumps and roots.

GSH. Glutahione (reduced), as tripeptide.

GST. Greenwich Standard Time.

Gulley. A small elongated depression in the land surface, usually dry except after a rainstorm. A large gulley is called a ravine.

GVWR. Gross vehicle weight rating.

H

H. See Hydrogen.

Habitat. The sum total of environmental conditions of a specific place that is occupied by an organism, a population or a community.

Half-Life. The time it takes certain materials, such as persistent pesticides or radioactive isotopes, to lose half their strength. For example, the half-life of DDT is 15 years; the half-life of radium is 1,580 years.

Halophyte. A plant that grows naturally in soils impregnated with salts.

Hammermill. A broad category of highspeed equipment that uses pivoted or fixed hammers or cutters to crush, grind, chip or shred solid wastes.

Hardpan. A dense hard layer in the subsoil, which obstructs penetration of roots and water. It consists either of soil particles so arranged and graded as to reduce materially the porosity of the soil, or more frequently, of soil which has been cemented by accumulations of chemicals formed by the continued precipitation of the dissolved salts of iron, calcium, magnesium, and other minerals, from water in a soil which lacks free drainage.

Hard Water. Water containing dissolved minerals such as calcium, iron and magnesium. The most notable characteristic of hard water is its inability to lather soap. Some pesticide chemicals will curdle or settle out when added to hard water.

HATREMS. Hazardous and Trace Substance Emissions System (EPA).

Hazardous Air Pollutant. According to law, a pollutant to which no ambient air quality standard is applicable and that may cause or contribute to an increase in mortality or in serious illness. For example, asbestos, beryllium and mercury have been declared hazardous air pollutants.

HC. Hydrocarbons.

HCL. Hydrochloric acid.

HDD. See Heating Degree Days.

HDPE. High Density Polyethylene.

HDV. Heavy duty vehicle.

Header. A large pipe or plenum installed to intercept the ends of a series of pipe – a manifold. A supply ditch for the irrigation of a field.

Heat, Available. The quantity of useful heat produced per unit of fuel if it is completely burned; the heat values of the dry flue gas and water vapor are deducted.

Heat Balance. An accounting of the distribution of the heat input and output of an incinerator, usually on an hourly basis.

Heat Exchanger. A device that transfers heat from one fluid to another without allowing them to mix.

Heat Island Effect. An air circulation problem peculiar to cities. Tall buildings, heat from pavements and concentrations of pollutants create a haze dome that prevents rising hot air from being cooled at its normal rate. A self-contained circulation system is put in motion that can be broken by relatively strong winds. If such winds are absent, the heat island can trap high concentrations of pollutants and present a serious health problem.

Heat Loss. The loss of heat transferred from a hot object by convection, conduction and radiation.

Heat of Combustion. The heat released, measure in BTU, when a unit quantity of waste or fuel is burned.

Heat Release Rate. The amount of heat liberated during complete combustion; it is usually expressed in BTU per hour per cubic foot of the internal volume of the furnace where the combustion takes place.

Heating Degree Days (HDD). A measure of the coldness of a region. To compute the HDD, average the high and low temperature of each day and add the resultant number of such days per year which are less than 65°F.

Heating Season. The coldest months of the year when pollution emissions are higher in some areas because of increased fossil fuel consumption.

Heavy Hydrogen. The hydrogen isotope having an atomic weight of 2 because of the addition of 1 neutron to the single proton in the ordinary hydrogen nucleus. Also called deuterium.

Heavy Metals. Metallic elements with high molecular weights, generally toxic in low concentrations to plant and animal life. Such metals are often residual in the environment and exhibit biological accumulation. Examples include mercury, chromium, cadmium, arsenic and lead.

Heavy Water. Water in which the molecule consists of a combination of heavy hydrogen and oxygen. It is used as a moderator in a nuclear reactor.

Herbicide. A pesticide chemical used to destroy or control the growth of weeds, bush and other undesirable plants. See pesticide.

Herbiciding. The use of chemicals for elimination or control of noxious or undesirable weeds or vegetation other than trees, shrubs, and desired plants.

Herbivore. An organism that feeds on vegetation.

HERL. Health Effects Research Laboratory – (EPA, Research Trinagle Park, N.C.).

Heterotrophic Organism. Organisms dependent on organic matter for food.

HEW. Department of Health, Education and Welfare (US).

Hg. See Mercury.

High Density Polyethylene. A material often used in the manufacture of plastic bottles that produces toxic fumes if incinerated.

High Volume Sampler. See Hi-volume sampler.

Histogram. A series of rectangles depicting data by frequency distribution and class interval heights usually plotted on an X-Y axis.

Hi-Vol. See Hi-Volume Sampler.

Hi-Volume Sampler. A device used in the measurement and analysis of suspended particulate pollution. Also called a Hi-Vol.

HIWAY. Highway Model, EPA.

H_2O. Water.

Hog Fuel Boiler. See bark boiler.

Hot. A colloquial term meaning highly radioactive.

Hot Blast Main. A duct-lined with refractory material, through which hot air passes from a hot blast stove to the bustle pipe of a blast furnace.

Hot Well. A reservoir for receiving warm condensed steam drawn from a condenser.

HR. Heart rate.

hr. Hour.

H_2S. Hydrogen sulfide.

H_2SO_4. Sulfuric acid.

Humidity. The percentage value of the amount of moisture (water vapor) in the atmosphere compared to that which could be under existing pressure and temperature conditions.

Humus. Decomposed organic material.

Hydration. The chemical process of combination or union of water with other substances.

Hydrator-Absorber. A single or double tower in which phosphorus pentoxide is hydrated to phosphoric acid and the resulting acid mist is absorbed.

Hydraulic Fly Ash Handling. A system using water-filled pipes or troughs in which fly ash is conveyed by means of gravity, water jets, or centrifugal pumps.

HYDRO. Hydrographic survey.

Hydrocarbons. A vast family of compounds containing carbon and hydrogen in various combinations, found especially in fossil fuels. Some hydrocarbons are major air pollutants, some may be carcinogenic and others contribute to photochemical smog.

Hydrocracking. A low-temperature catalytic method of converting crude oil, residual oil, petroleum tar, and asphalt to high octane gasoline, jet fuel, and/or high-grade fuel oil. The process combines cracking, hydrogenation, and isomerization.

Hydrodesulfurization. A desulfurization process in which the oil is heated with hydrogen.

Hydroelectric Energy. The energy derived from the head or potential of water flowing to the seas.

Hydrogen Sulfide (H_2S). A malodorous gas made up of hydrogen and sulfur with the characteristic odor of rotten eggs. It is emitted in the natural decomposition of organic matter and is also the natural accompaniment of advanced stages of eutrophication. H_2S is also a byproduct of refinery activity and the combustion of oil during power plant operations. In heavy concentrations, it can cause illness.

Hydrogenation. The chemical addition of hydrogen to a material at high pressure in the presence of a catalyst.

Hydrology. The science dealing with the properties, distribution and circulation of water and snow.

Hydrolysis. A change in the chemical composition of matter produced by combination with water. Sometimes loosely applied in sewage practice with reference to the liquefaction of solids in a tank as a result of bio-chemical action.

Hydrometallurgy. The treatment of ores, concentrates, and other metal-bearing materials by wet processes, usually involving the solution of some component, and its subsequent recovery from the solution.

Hydrophyte. A plant which grows naturally in water, or in saturated soils.

Hydrotreating. A treating process using hydrogen for the desulfurization of cracked distillates.

Hydrometer. An instrument for determining the dew point.

Hygroscopic Water. Water molecules held on surfaces of particles by forces of adhesion.

Hygrothermograph. An instrument that records both humidity and temperature on the same chart.

Hyperbolic Natural Draft. A large hyperbolic shaped tower that uses air flow established by the natural draft (chimney effect) to remove large amounts of heat. Used primarily in connection with large power plants.

Hypertrophic. Waters having high nutrient content.

Hz. Hertz (frequency).

I

I. See Iodine.

IACOMS. International Advisory Committee on Marine Sciences.

IAPO. International Association of Physical Oceanography.

IC. Weather abbreviation for "ice crystals."

ICG. Weather abbreviation for "icing."

ICGIP. Weather abbreviation for "icing in precipitation."

ID. Identification.

IEEE. Institute of Electrical and Electronic Engineers.

IF. Weather abbreviation for "ice fog."

Ig. Immunoglobulins.

IgA. Alpha globulin.

IgG. Gamma globulin.

IGOSS. Integrated Global Ocean Station System; Scientific study of relationship between ocean and atmosphere including marine pollution monitoring, under auspices of World Meterological Organization.

Il. See Illinium.

Illinium (Il). A rare earth element with an estimated atomic weight of 146 and an atomic number of 61.

IM. Inspection and Maintenance, EPA.

Imhoff Tank. A deep two-storied tank with an upper continuous flow sedimentation chamber and a lower sludge-digestion chamber.

Impactor. A sampling device that employs the principle of impaction (impingement). The cascade impactor is a specific instrument that employs several impactions in series to collect smaller sizes of particles.

Impedance. The rate at which a substance can absorb and transmit sound; also in alternating-current electrical circuits the total opposition offered to current. It is comprised of resistance, inductive and capacitive reactance.

Impingement. In air sampling, impingement refers to a process for the collection of particulate matter in which the gas being sampled is directed forcibly against a surface. 1. Dry impingement: The process of impingement in the gas stream where particulate matter is retained upon the surface against which the stream is directed. The collecting surface may be treated with a film of adhesive. 2. Wet impingement: The process of impingement in a liquid which retains the particulate matter.

Impingement Separators. Devices using the principle that when a gas stream carrying particulate matter impinges on a body, the gas is deflected around the body, while the particles, because of their greater inertia, tend to strike the body and be collected on its surface. The bodies may be in the form of plates, cylinders, ribbons or spheres.

Implementation Plan. A document of the steps to be taken to ensure attainment of environmental quality standards within a specified time period. Implementation plans are required by various laws.

Impoundment. A body of water, such as a pond, confined by a dam, dike, floodgate or other barrier.

In. See Indium.

in. Inch.

Incidence Angle. See angle of incidence.

Incineration. The controlled process by which solid, liquid or gaseous combustible wastes are burned and changed into gases; the residue produced contains little or no combustible materials.

Incinerator. An engineered apparatus used to burn waste substances and in which all the combustion factors – temperature, retention time, turbulence and combustion air – can be controlled.

Indium (In). A rare metallic element discovered in zinc ores. It is a very soft silvery metal whose atomic weight is 114.76 and atomic number is 49.

Induced Draft. The negative pressure created by the action of a fan, blower, or ejector located between an incinerator and a stack.

Induced Draft Fan. A fan that exhausts hot gases from heat-absorbing equipment, dust collectors or scrubbers.

Inert Gas. A gas that does not react with other substances under ordinary conditions.

Inertial Separator. An air pollution control device that uses the principle of inertia to remove particulate matter from a stream of air or gas.

Infiltration. The flow of a fluid into a substance through pores or small openings. Commonly used in hydrology to denote the flow of water into soil material.

Infiltration Air. Air that leaks into the chambers or ducts of an incinerator.

Influent. Liquid flowing into a waste water treatment plant or other process plant.

Inoculum. Material such as bacteria placed in compost or other medium to initiate biological action.

Insolation. Incoming solar radiation reaching the earth's surface.

Instantaneous Sampling (Grab Sampling). Obtaining a sample in a very short period of time, such that this sampling time is insignificant in comparison with the duration of the operation or the period being sampled.

Insulation. Non-conducting materials which are used to reduce the rate of energy transfer through barriers, walls, ceilings, and floors. Used extensively in solar energy systems.

Integrated Pest Control. A system of managing pests by using biological, cultural and chemical means.

Interceptor Sewers. Sewers used to collect the flows from main and trunk sewers and carry them to a central point for treatment and discharge. In a combined sewer system, where street runoff from rains is allowed to enter the system along with sewage, interceptor sewers allow some of the sewage to flow untreated directly into the receiving stream, to prevent the plant from being overloaded.

Intermittent Sampling. Sampling successively for limited periods of time throughout an operation or for a pre-determined period of time. The durations of sampling periods and of the intervals in between are not necessarily regular and are not specific.

Internal-Combustion Engine. An engine in which both the heat energy and the ensuing mechanical energy are produced inside the engine proper.

Interruptible Gas. Gas sold whereby the seller may curtail or stop delivery, generally at his option. The gas customer under these conditions is expected to have standby equipment capable of taking over 100% of his needs by an alternate fuel.

Interstate Carrier Water Supply. A water supply whose water may be used for drinking or cooking purposes aboard common carriers (planes, trains, buses and ships) operating interstate. Interstate carrier water supplies are regulated by the Federal government.

Interstate Waters. According to law, waters defined as: (1) rivers, lakes, and other waters that flow across or form a part of State or international boundaries; (2) waters of the Great Lakes; (3) coastal waters – whose scope has been defined to include ocean waters seaward to the territorial limits and waters along the coastline (including inland streams) influenced by the tide.

Interstitial Water. Water contained in the interstices or voids of a surface.

Inversion. An atmospheric condition where a layer of cool air is trapped by a layer of warm air so that it cannot rise. Inversions spread polluted air horizontally rather than vertically so that contaminating substances cannot be widely dispersed. An inversion of several days can cause an air pollution episode.

Inversion Base. The lowest height in the atmosphere at which the temperature ceases to decrease with height.

Invert. The floor, bottom, or lowest portion of the internal cross section of a closed conduit. Used particularly with reference to aqueducts, sewers, tunnels, and drains.

I/O. Input/output – a computer peripheral unit.

Iodine (I). A grayish black lustrous solid element with an atomic weight of 126.92 and an atomic number of 52. Forms a blue-violet gas with an irritating odor and a brown solution with water.

IOF. International Oceanographic Foundation, Miami, Fl.

IOI. International Ozone Institute.

Ion. An electrically charged atom or group of atoms. A negatively charged atom is called a negative ion; a positively charged one a positive ion. See ionization.

Ion Exchange. A reversible exchange of ions (electrically charged atoms or groups of atoms) between a solid and a liquid in water and waste water treatment, the solids, ion-exchange resins, exchange the polluting ions in the water for ions originally on the resin.

Ionization. The process by which an atom becomes electrically charged, by the removal or addition of one or more of its extranuclear electrons. So that the electrical balance between the electrons and the protons within the atom's nucleus is destroyed. An atom with more than its normal complement of electrons has a negative charge; with less, it has a positive one.

Ionization Chamber. A device roughly similar to a Geiger counter that reveals the presence of ionizing radiation.

IPFPE. Industrial Process Fugitive Particulate Emissions – particulate matter which escapes from a defined process flow stream due to leakage, materials charging/handling, inadequate operational control, lack of reasonably available control technology, transfer or storage.

Ir. See Iridium.

Iridium (Ir). A very hard metallic element whose atomic weight is 193.1 and atomic number is 77. It is used in alloys for hardening and to make weights and measures.

Iron (Fe). A tough greyish white metal element whose atomic weight is 55.84 and atomic number is 26. Used extensively as a metal since prehistoric times.

Isobars. Lines of equal barometric pressure on a chart.

Isobath. A line connecting all points which are the same vertical distance above the upper or lower surface of a water bearing formation or aquifer.

Isohyet. A line on the earth's surface as represented on a chart connecting all points of equal precipitation.

Isomerization. A reaction which alters the fundamental arrangement of the atoms in a molecule without adding or removing anything from the original material. In the petroleum industry, straight-chain hydrocarbons are converted catalytically to branched-chain hydrocarbons of substantially higher octane number by isomerization.

Isopleths. Lines of equal intensity of pollutants on a chart.

Isotope. A variation of an element having the same atomic number as the element itself, but having a different atomic weight because of a different number of neutrons. Different isotopes of the same element have different radioactive behavior.

ISR. Indirect source review.

J

J. Joule.

Jackson Turbidity Units (JTU). A measure of the turbidity of water, proportional to the ppm silica, where 100 ppm silica equals 21.5 JTU.

Jet. A nonbuoyant plume.

Jetsam. Debris and loose material deliberately jettisoned and floating on the surface of waters.

Jetting. A method of well sinking or pile driving using high pressure water through a jet nozzle.

Jig. A device which separates coal from foreign matter by means of their difference in specific gravity in a water medium. The water pulsates up and down causing the heavy material to work to the bottom. Also, a mechanical positioner or spacer.

JTU. See Jackson Turbidity Units.

K

K. Weather abbreviation for "smoke;" also, signifies one thousand; also see potassium.

°K. Degree Kelvin.

Kelly. Uppermost section of the column of drill pipe. It passes through and engages in the opening in the rotary table. The outer shape may be square, hexogonal, or round with lengthwise grooves or flutes cut into the outside wall. Used in well construction.

Kerogen. Bituminous material occurring in shale and yielding oil when heated.

Kettle. An open-top vessel used in carrying out metallurgical operations on low-melting-point metals; for example, in drossing and desilverizing lead. An open or (usually) closed vessel for preparing paints, varnishes, and resins.

KEV. Kilo-electron-volts.

Kg. Kilogram.

Kg N/yr. Kilograms nitrogen per year.

KH_z. Kilohertz (kilocycles per sec.).

Kiln. A furnace in which the heating operations do not involve fusion. Kilns are most frequently used for calcining, and free access of air is permitted. The raw materials may be heated by the combustion of solid fuel with which they are mixed, but more usually they are heated by gas or the waste heat from other furnaces.

Kiln Gas. Hot effluent gases from a kiln. Unless controlled, these gases can be the largest source of particulates in a plant.

Kilo. Thousand.

Kilowatt. A unit of power equal to 1000 watts.

Kilowatt-hour (KWH). A unit of electrical work or energy equal that expended by one kilowatt in one hour.

Kinematic Viscosity. The relative resistance of an air mass to the diffusion of pollutant emissions.

Kinetic Energy. Energy possessed by a mass because of its motion, e.g. the falling water of a dam has the energy to put turbines into motion.

Kinetic Theory. Holds that a material body is not one continuous, uninterrupted extent but is made up of minute, invisible particles – molecules – that are in constant motion, oscillating, bumping into each other, and bouncing back. Increasing the temperature causes a rise in the speed of the molecules, thus increasing the kinetic energy of the molecules.

Klystron. An electron tube in which bunching of electrons is produced by electric fields and which is used for the generation and amplification of ultra-high frequency current.

km. Kilometer.

Knockout Drum. A drum or vessel constructed with baffles through which a mixture of gas and liquid is passed to disengage one from the other. As the mixture comes in contact with the baffles, the impact frees the gases and allows them to pass overhead; the heavier substance falls to the bottom of the drum.

Kr. See Krypton.

Kraft. A strong paper or board made from wood pulp derived from wood chips boiled in an alkaline solution containing sodium sulfate.

Krypton (Kr). An inert rare gaseous element occurring in very small amounts in the atmosphere. Its atomic weight is 83.7 and atomic number is 36.

kt. Knot, nautical mile per hour.

KWH. See kilowatt hour.

Kymogram. A record made by a Kymograph.

Kymograph. An instrument which graphically records motion or pressure.

L

l. Liter.

$L_{1/3}$. Noise level exceeded 1/3 of the time.

L_{10}. Noise level exceeded 10 per cent of the time.

L_{50}. Noise level exceeded 50 per cent of the time.

L_{90}. Noise level exceeded 90 per cent of the time.

La. See Lanthanum.

LACT. Latest Available Control Technology.

LAER. Latest Available Emission Reduction.

Lagoon. In waste water treatment, a shallow pond usually man-made where sunlight, bacterial action and oxygen interact to restore waste water to a reasonable state of purity.

Langley. A unit for measuring solar radiation, equal to one calorie of radiation energy per square centimeter. Most meteorological stations report solar radiation in terms of total Langley's received on a horizontal surface at ground level. One Langley of solar radiation per hour is equivalent to 3.69 $BTU/ft^2/hr$.

Lanthanum (LA). A metallic rare earth element which resembles iron in its physical properties. Its atomic weight is 138.92 and its atomic number is 57.

Lantz Process. A destructive distillation technique in which the combustible components of solid waste are converted into combustible gases, charcoal, and a variety of distillates.

Lapse Rate. The rate at which temperature drops with increasing altitude; the negative of the vertical temperature gradient.

Larvicide. A chemical used to destroy or control undesirable larva such as mosquito larva.

Larviciding. The elimination or control of larval arthropods by the use of chemicals, such as FLIT-MLO for control of mosquito larvae.

Lat. Latitude.

Lateral Sewers. Pipes running underneath city streets that collect sewage from homes or businesses.

Launder. A trough, channel, or gutter usually of wood, by which water is conveyed. Specifically, in mining, a chute or trough for conveying powered ore, or for carrying water to or from the crushing apparatus.

Lawrencium (Lr). Radioactive element. Atomic Weight (longest lived known isotope) 260; atomic number 103.

lb. Pound.

LC_{50}. Median lethal concentration, a standard measure of toxicity. LC_{50} indicates the concentration of a substance that will kill 50 percent of a group of experimental insects or animals.

LCT. Local civil time.

LD_{50}. Lethal dose 50%; dose which is lethal to 50% of subjects.

LDH. Lactate dehydrogenase.

L_{dn}. Day night equivalent sound level.

LDV. Light duty vehicle.

Leachate. Liquid that has percolated through solid waste or other mediums and has extracted dissolved or suspended materials from it.

Leaching. The process by which soluble materials in the soil, such as nutrients, pesticide chemicals or contaminants, are washed into a lower layer of soil or are dissolved and carried away by water.

Lead (Pb). A very soft, highly malleable, blue-white lustrous metal element whose atomic weight is 207.22 and atomic number is 82. Used extensively since biblical times in a variety of purposes and compounds.

L_{eq}. Sound level which has the same amount of sound energy over a given time period as the actual time varying sound level. Also called equivalent energy level.

Li. See Lithium.

Life Cycle. The phases, changes or stages an organism passes through during its lifetime.

Lift. In a sanitary landfill, a compacted layer of solid waste and the top layer of cover material.

Lignite Coal (Brown Coal). A brownish-black variety of coal, usually high in moisture and low in Btu's. Lignite is one of the earlier stages in the formation of bituminous coal.

Limnology. The study of the physical, chemical, meteorological and biological aspects of fresh waters.

Line Source. A source of air pollutants as may be depicted in a diffusion model by a straight line configuration. Examples are streets, highways, closely-spaced multiple stacks, airport runways and aircraft flight paths.

Lining. The material used on the inside of a furnace wall; usually a high-grade refractory tile or brick or a plastic refractory material.

Liquefaction. The changing of the organic matter in sewage from an insoluble to a soluble state, and effecting a reduction in its solid contents.

Liquid Concentrator. A device for using evaporation to reduce liquid content of a mixture, leaving a higher percentage of solids, usually greater than 50%.

LIRAQ. Livermore Regional Air Quality Model (Photochemical oxidant concentration model uses HC measurements).

Lithium (Li). A soft white alkaline metal element whose atomic weight is 6.940 and atomic number is 3. It is used in medicine.

Lithosphere. That part of the earth which is composed predominantly of rocks and including the disintegrated rock materials known as soils and subsoils together with everything in this rocky crust.

Littoral. Related to coastline; or more specifically between high and low water marks.

L_n. Noise level exceeded, n per cent of the time.

Lofting. A term describing plume rise where more vertical diffusion takes place than in horizontal plane.

Long. Longitude.

Long-Wave Radiation. Electromagnetic radiation that is of less energy and longer wave length than visible light – primarily with wave lengths greater than 3 microns.

Looping. A term describing a plume rise which forms loops in vertical plane because of alternate thermals and sinks.

lpm. Liter per minute.

LPS. Bacterial lipopolysaccharide.

LOX. Liquid Oxygen.

Lr. Lawrencium.

LRC. Landing Rate Count (arthropod control).

LRSF. Long range systems forecast (weather).

LST. Local Standard Time.

LT_{50}. Median lethal time.

Lu. See Lutecium.

Lutecium (Lu). A rare mineral element whose atomic weight is 175.0 and atomic number is 71.

Lysimeter. An instrument to measure water percolation rate in soils – may also be used to collect percolated water for quality analysis.

LZT. Local zone time.

M

m. Meter.

m^3. Cubic meter.

Ma. See Masurium.

Machine. A device which uses mechanical energy to get work done.

Magnesium (Mg). A light weight white tough metal element whose atomic weight is 24.32 and atomic number is 12. It is used in photography, medicines, and in alloys for lightness.

Malfunction. Any unavoidable mechanical and/or electrical failure of air pollution control equipment or process equipment or of a process to operate in a normal or usual manner.

Manganese (Mn). A hard grey-white metal element whose atomic number is 54.93 and atomic number is 25. It is used in alloys with copper, brass and nickel for hardening.

Manifold. A pipe or header for collecting a fluid or gas from or distributing a fluid or gas to, a number of pipes or tubes.

Manometer. A u-shaped tube device used for measuring fluid pressures.

Manufactured Gas. Fuel gas manufactured from coal, oil, etc., as differentiated from natural gas.

Marsh. A low-lying tract of soft, wet land that provides an important ecosystem for a variety of plant and animal life but often is destroyed by dredging and filling.

Masking. Covering over of one sound or element by another. Quantitatively, masking is the amount the audibility threshold of one sound is raised by the presence of a second masking sound. Also used in regard to odors.

Masurium (Ma). A mineral element whose atomic weight is 43. It is very rare.

Material Balance. An accounting of the weights of material entering and leaving a process.

Matte. A metallic sulfide mixture formed in smelting sulfide ores of copper, lead, and nickel.

Max. Maximum.

MCA. Manufacturing Chemists Association.

MCF. Thousands of cubic feet.

MCL. Maximum Containment Level (EPA Drinking Water Regulations).

Md. Mendelevium.

Mechanical, Centrifugal Separators. A device for separating particulates. A rotating fan blade exerts a large centrifugal force on the particulates, ejecting them from the tips of the blades to a skimmer bypass leading into a dust hopper.

Mechanical Crossflow Tower. A cooling tower using induced or forced-air fans to draw air horizontally across the fill.

Mechanical Draft. A type of cooling tower in which fans are used to induce or force air through the tower.

Mechanical Energy. Energy in a form which can do work directly.

Mechanical Scrubber. A scrubber in which the water spray is generated by a rotating element or disk (see also scrubber).

Mechanical Turbulence. The erratic movement of air caused by local obstructions such as buildings.

mega. A prefix meaning 1 million.

Membrane Barrier. Thin layer of material impermeable to the flow of gas or water.

Membrane Filter. Controlled pore filters commonly composed of cellulose esters. They can be manufactured with uniformly controlled pore size. Nylon mesh may be used for reinforcement. Types commonly used for air sampling have a pore size of about 0.45 to 0.8 microns. The pores constitute 80 to 85 percent of the filter volume. Because of electrostatic forces and the formation of a precoat of collected particles on the surface, these filters can collect down to about 0.1 microns in diameter.

Mendelevium (Md). Man-made radio-active element. Atomic weight (most stable known isotope) 258; atomic number 101.

Meq. Milliequivalent.

Mercaptans. Organic compounds having the general formula R-SH (where R represents any hydrocarbon radical) which are analogous to the alcohols and phenols but which contain sulfur in place of oxygen. The simpler mercaptans have strong, repulsive odors.

Mercury (Hg). A heavy silver-white shining liquid metal whose atomic weight is 200.61 and atomic number is 80. The element and its more important salts are used in a variety of purposes. Mercury is highly toxic if breathed or ingested. Mercury is residual in the environment, showing biological accumulation in all aquatic organisms, especially fish and shell fish. Chronic exposure to airborne mercury can have serious effects on the central nervous system.

MERL. Municipal Environmental Research Laboratory (EPA-Cincinnati).

Mesh. The number of holes per linear unit in a sieve or gauze, or the space between the wires of the sieve expressed in inches or millimeters.

Mesophyte. A plant that grows under medium or usual conditions of atmospheric moisture supply, as distinguished from one which grows under dry or desert or very wet conditions.

Methane. Colorless, nonpoisonous and flammable gaseous hydrocarbon. Methane, (CH_4) is emitted by marshes and by dumps undergoing anaerobic decomposition.

Metric Ton. 2204.6 pounds or 1000 kilograms.

MG. Megagram.

Mg. See magnesium.

Mg. Milligram.

MGD. Million Gallons Per Day.

mg/1. Milligrams per liter.

Mho. Electrical unit of conductance; reciprocal of ohm.

MHz. Megahertz (frequency).

micro. A prefix meaning 1/1,000,000 abbreviated by the Greek letter μ.

Microbes. Minute plant or animal life. Some diesase-causing microbes exist in sewage.

Micron. A unit of length, usually used to express the dimensions of very small particles, and hence, to rate filters. One micron equals one millionth of a meter, or 0.00003937 in.

Microorganism. Minute organism either plant or animal, invisible or barely visible to the naked eye.

milli. A prefix meaning 1/1,000.

Milligrams per Liter (Mg/1). Sometimes used to express contaminant concentrations in gases. One pound equals 454 grams or 454,000 milligrams; one cubic foot equals 28.32 liters; hence, 1 mg/liter is approximately equivalent to 62.4 pounds per million cubic feet.

Min. Minute; minimum.

Mist. Liquid particles in air formed by condensation of vaporized liquids. Mist particles vary from 500 to 40 microns in size. By comparison, fog particles are smaller than 40 microns in size.

Mixed Liquor. A mixture of activated sludge and water containing organic matter undergoing activated sludge treatment in the aeration tanks.

Mixed-Media Gravity Filter. A filter using more than one filtering media (such as coal and sand).

Mixing Chamber. A chamber usually placed between the primary and secondary treatment tanks of a wastewater plant.

Mixing Depth. The expanse in which air rises from the earth and mixes with the air above it until it meets air equal or warmer in temperature.

Mixing Height. Heights to which a pollutant (air) can be expected to mix vertically as determined by the lapse rate and/or turbulence.

ml. Milliliter.

mM. Millimoles.

MMT. Methylcyclopentadienyl Manganese Tri-carbonyl (manganese additive to raise gasoline octane level).

Mn. See manganese.

Mo. See Molybdenum.

Mobile Source. A moving source of air pollution such as an automobile.

Model/Modeling. A mathematical or physical representation of an observable situation. In air pollution control, models afford the ability to predict pollutant distribution or dispersion from identified sources for specified weather conditions.

Moderator. A material that reacts very little to neutron bombardment. Used in nuclear reactors to slow down the neutrons for greater absorption by the fissionable fuel.

Modified Coal. Coal of a stoker size containing a controlled percentage of fines.

Molecule. The smallest part of a substance that can exist separately and still retain its chemical properties and characteristic composition; the smallest combination of atoms that will form a given chemical compound.

mol. Mole.

mol. wt. Molecular weight.

Molybdenum (Mo). Very hard silver-white metal whose atomic weight is 96.0 and atomic number is 42. It is used as a hardening alloy.

Monitoring. Periodic or continuous determination of the amount of pollutants or radioactive contamination present in the environment.

Monolithic Lining. A refractory lining or construction made in large sections on site: Conventional layers and joints of brick construction are not used.

Moraine. The large mass of drift and other material transported by a glacier and deposited by it in the process of melting.

Motor. An engine; also, a device for changing electrical energy into mechanical energy.

MPC. Maximum permissible concentration.

MPN. Most probable number; the number of organisms per unit volume in bacterial analysis by dilution in accordance with statistical theory.

MSD. Marine Sanitation Device.

Muck Soils. Soils made from decaying plant materials.

Mucus. The sticky fluid covering the airway linings of the respiratory system.

Mulch. A layer of wood chips, dry leaves, straw, hay, plastic strips or other material placed on the soil around plants to retain moisture, to prevent weeds from growing and to enrich soil.

Multicyclone (also Multiple Cyclone or Multiclone). A dust collector consisting of a number of cyclones, operating in parallel, through which the volume and velocity of gas can be regulated by means of dampers to maintain dust-collector efficiency over the load range.

Multiple-Chamber Incinerator. Any incinerator consisting of a primary combustion chamber, mixing chamber, and secondary combustion chamber in series. The chambers are separated by refractory walls, and interconnected by gas passage ports.

Multiple-Hearth Type Roaster. See roasting furnace.

Municipal Incinerator. An incinerator owned or operated by government or by a person who provides incinerator service to government or others; a device designed for and used to burn waste materials of any and all types.

N

N. Newton.

N_2. See Nitrogen.

Na. See Sodium.

NAAQS. National Ambient Air Quality Standards – A legal limit on the level of atmospheric contamination necessary to protect against adverse effects on public health and welfare. Primary standards are those related to health effects, and secondary standards are related to protection against adverse welfare effects.

NACO. National Association of Counties.

NADB. National Air Data Branch; National Aerometric Data Bank.

NADH. Nicotinamide adenine dinucleotide (reduced).

NADPH. Nicotinamide adenine dinucleotide phosphate (reduced); a substance involved in biochemical reactions.

NAM. National Association of Manufacturers.

NAMF. National Association of Metal Finishers.

Nanosec. 1 billionth of a second of time.

NAPA. National Asphalt Pavement Association.

Nappe. The sheet or curtain of water freely overflowing a weir or dam.

NAQTS. National Air Quality Trends Stations (USEPA).

NARC. National Association of Regional Councils.

NARI. National Association of Recycling Industries.

NAS. National Academy of Sciences.

NASA. National Aeronautics and Space Administration.

NASN. National Aerometric Surveillance Network.

National Ambient Air Quality Standards (NAAQS). See NAAQS.

Natural Draft. The negative pressure created by the height of a stack or chimney and the difference in temperature between the gases and the atmosphere.

Natural Gas. A fuel gas occurring naturally in certain geologic formation. Natural gas is usually a combustible mixture of methane and hydrocarbons.

Natural Selection. The natural process by which the organisms best adapted to their environment survive and those less well adapted are eliminated.

Navigable Waters. Those waters capable of supporting boat or ship traffic as declared by the federal agency authorized by Congress to make such determination.

Nb. See columbium.

NBS. National Bureau of Standards.

Nd. See neodymium.

NE. Northeast.

Ne. See Neon.

Neat Cement Slurry. A mixture of cement and water (usually one bag to 5–7 gallons of water) used in well construction or modification.

Necrosis. Death of plant cells resulting in a discolored, sunken area or death of the entire plant.

NEDS. National Emissions Data System (EPA).

NEDSXREF. Number providing cross reference between NEDS and enforcement data systems.

NEF. Noise exposure forecast.

Neodymium (Nd). A rare earth metallic element whose atomic weight is 144.27 and atomic number is 60. It forms a pink salt.

Neon (Ne). A gaseous element present in small quantities in the atmosphere whose atomic weight is 20.183 and atomic number is 10. Its principal use is in lighting fixtures since it glows in a vacuum.

NEPA. National Environmental Protection Act.

Nephalometer. An instrument which uses light rays to measure the opacity and hence the turbidity of water.

Neptunium (Np). An element. Atomic weight (most stable isotope) 237; atomic number 93.

NESHAPS. National Emission Standards for Hazardous Air Pollutants.

Net Ton. 2000 pounds (sometimes known as a "short ton").

Neutral. In hydrostatic equilibrium. A neutral atmosphere is characterized by an adiabatic lapse rate, i.e., by potential temperature constant with height.

Neutron. See atom.

Neve. A granular mass of a snow field which is somewhat between clear ice and snow.

New Source. Any source other than an existing source. New sources include reactivating existing sources which have been shut down for a period of more than one year.

NFMA. National Forest Management Act.

ng. Nanogram.

NHTSA. National Highway Traffic Safety Administration.

Ni. See Nickel.

Nickel (N_i). A hard, malleable, ductile, conducting, magnetic, metal element whose atomic weight is 58.69 and atomic number is 28. It has many common uses when alloyed with copper, steel, etc.

NIEHS. National Institute of Environmental Health Sciences.

NIH. National Institute of Health, Bethesda, Md.

Niobium (Nb). See Columbium.

NIOSH. National Institute of Occupational Safety and Health.

Nitric Oxide (NO). A gas formed in great part from atmospheric nitrogen and oxygen when combustion takes place under high temperature and high pressure, as in internal combustion engines. NO is not itself a pollutant; however, in the ambient air, it converts to nitrogen dioxide, a major contributor to photochemical smog.

Nitrification. The oxidation of organic nitrogen intc nitrates through biochemical action.

Nitrogen (N). A gaseous element which makes up about four-fifths of the earth's atmosphere. It has an atomic weight of 14.008 and an atomic number of 7. It occurs in all living things and compounds containing nitrogen have many uses.

Nitrogen Dioxide (NO_2). A compound produced by the oxidation of nitric oxide in the atmosphere; a major contributor to photochemical smog.

Nitrogen Oxides. A general term pertaining to a mixture of nitric oxide (NO) and nitrogen dioxide (NO_2).

Nitrogenous Wastes. Wastes of animal or plant origin that contain a significant concentration of nitrogen.

ng. Nanogram.

Nm. Nanometer.

NMHC. Non-methane hydrocarbons.

NMMSA. Non Metallic Mine Safety Act.

NMOC. Non-methane organic materials.

No. Nobelium.

NO_2. Nitrogen Dioxide.

NOAA. National Oceanic and Atmospheric Administration; U.S. Department of Commerce.

Nobelium (N_o). An element. Atomic weight (longest-lived isotope) 259; atomic number 102.

Nocturnal Radiation. The radiation or loss of heat at night from a warm object such as a house to the cold sky and surroundings. Also applied to bodies of water which do not cool as rapidly as adjacent land areas at night. See heat island effect.

NOE's. Notice of Effectiveness of Prohibition Orders (FEA).

NOHb. Nitrosohemoglobin.

NOI. Notice of Intention.

Noise. Any undesired audible signal. Thus, in acoustics, noise is any undesired sound.

NO_x. Oxides of nitrogen.

Np. Neptunium.

NPA. Non-passenger automobiles.

NPAAQS. National Pollution Air attainment Quality Standards (EPA).

NPDES. National Pollutant Discharge Elimination System (USEPA).

NRC. National Resources Council.

N.S. Not statistically significant.

NSIS. National Source Inventory Section (EPA).

NSPS. New Source Performance Standards.

NSR. New Source Review, EPA.

NTA. Nitrilotriacetic acid, a compound once used to replace phosphates in detergents.

NTOC. National Training and Operational Technology Center operated by EPA Cincinnati, Ohio.

NTSA. National Trails System Act.

NTU. Nephalometer turbidity units.

Nuclear Energy. The force released by nuclear decay; radioactivity.

Nuclear Power Plant. Any device, machine or assembly that converts nuclear energy into some form of useful power, such as mechanical or electrical power. In a nuclear electric power plant, heat produced by a reactor is generally used to make steam to drive a turbine that in turn drives an electric generator.

Nuclear Reactor. See Reactor.

Nucleon. Any constituent particle of the nucleus. At present, it is applied to protons and nuetrons, but it will include any other particle found to exist in the nucleus.

Nuclide. Any species of atom that exists for a measurable period of time whose nuclear structure is distinct from that of any other species. Thus, each isotope of an element is a separate nuclide.

Nuisance. Whatever is injurious to health, indecent, or offensive to the senses, or an obstruction to the free use of property, so as essentially to interfere with the comfortable enjoyment of life or property.

Nutrients. Elements or compounds essential as raw materials for organism growth and development; for example, carbon, oxygen, nitrogen and phosphorus.

NWWA. National Water Well Association; Worthington, Ohio; also National Water Works Association, Denver, Col.

O

O. See oxygen.

O_2. See oxygen.

O_3. See ozone.

OAP. Office of Air Programs (EPA).

OAQPS. Office of Air Quality Performance Standards, EPA.

OAQS. Office of Air Quality Standards (EPA).

OAWM. Office of Air and Waste Management (EPA).

OBA. Oxygen breathing apparatus.

Objectionable Odor. Any odor present in the atmosphere which by itself or in combination with other odors, is or may be harmful or injurious to human health or welfare, which unreasonably interferes with the comfortable use and enjoyment of life or property or which creates a nuisance.

Obnoxious Emission. Any emission which causes or produces irritation or discomfort to the eyes, ears, nose, skin or respiratory passages.

OBSC. Weather abbreviation for "obscure."

Ocean Disposal. The depositing of waste into ocean or estuarine waters.

Ocean Thermal Gradients. An energy source based on differential temperatures; warm surface and cool deep waters; tropical ocean areas may have 85°F at surface and 35°F at depths less than 2000 ft.

OCS. Outer Continental Shelf.

OCZM. Office of Coastal Zone Management (NOAA).

Odor. A sensation resulting from stimulation of the olefactory organ.

Odorant. A gaseous nuisance that is offensive or objectionable to the smell.

Odor Counteractant. A selected odor-emitting material whose odor combined with an objectionable odor creates a neutral, indistinguishable, or acceptable odor.

Odor-Intensity. The numerical or verbal indication of the strength of an odor.

Odor Masker. A selected acceptable odor emitting material used to offset or overpower an objectionable odor.

Odor Pervasiveness. The ability of an odor to diffuse into a large volume of air and still continue to possess a detectable intensity. A pervasive odor is one whose odor intensity changes very little on dilution.

Odor Quality. A verbal description of an odor. The quality may be described in terms of such familiar odorants as coffee, onions, lemons, or by associating an unfamiliar odor with a familiar odor.

Odor Threshold. The lowest concentration of an odor in air that can be detected by a human.

Odor Units. That quantity of odor necessary to contaminate one cubic foot of air to threshold or barely perceptible level. The number of odor units is equal to the volumes (scf) of air necessary to dilute the concentration of odorant in one volume (scf) of air to the threshold concentration.

OEA. Office of Environmental Affairs (USDAC).

OEMI. Office of Energy, Minerals and Industry (EPA-R&D).

Ogee. A reverse curve shaped like an elongated –S.

OHEE. Office of Health and Ecological Effects (EPA).

Ohm. Electrical unit of resistance in a circuit in which a potential of one volt produces one ampere of current flow.

Oil Burner. Any device for the introduction of vaporized or atomized fuel oil into a furnace.

Oil-Effluent Water Separator. Any tank, box, sump, or other container in which any petroleum product entrained in water is physically separated and removed prior to out-fall, drainage, or recovery of the water.

Oil Spill. The accidental discharge of oil into oceans, bays or inland waterways. Methods of oil spill control include chemical dispersion, combustion, mechanical containment and absorption.

Oiticica (oil). A drying oil obtained from the kernels of the fruit of the oiticica tree that is similar to tung oil in many properties and is used chiefly in varnishes, paints, and printing inks.

Oleoresin. A varnish or paint vehicle, made of plant oils and resins, usually cooked.

Oleum (Fuming Sulfuric Acid). A heavy, oily, strongly corrosive liquid that consists of a solution of sulfur trioxide in anhydrous sulfuric acid. It fumes in moist air and reacts violently with water.

Oligotrophic Lakes. Deep lakes that have a low supply of nutrients and thus contain little organic matter. Such lakes are characterized by high water transparency and high dissolved oxygen.

OMB. Office of Management and Budget.

ONAC. Office of Noise Abatement and Control (EPA).

Onstream Time. The length of time a unit is in actual production.

Opacity. Degree of obscuration of light. For example, a window has zero opacity; a wall is 100 per cent opaque. The Ringelmann system of evaluating smoke density is based on opacity.

Open Burning. Uncontrolled burning of wastes in an open dump.

Open Dump. See dump.

Open Hearth Furnace. Reverberatory furnace, containing a basin-shaped hearth, for melting and refining suitable types of pig iron, iron ore, and scrap for steel production.

ORD. Office of Research and Development (EPA).

Ore and Lime Boil. Reactions which occur in an open hearth furnace when carbon monoxide is produced by the oxidation of carbon. Ore boil is a violent agitation of the metal as it escapes during this process; lime boil occurs when the limestone decomposes and the carbon dioxide gas escapes. The second reaction begins before the first is completed.

Organic. Referring to or derived from living organisms. In chemistry, any compound containing carbon.

Organic Solvents. See volatile organic compounds.

Organic Sulfur. The difference between the total sulfur in coal and the sum of the pyritic sulfur and sulfate sulfur.

Organism. Any living human, plant or animal.

Organoleptic. Affecting or making an impression upon one or more of the sense organs.

Organophosphates. A group of pesticide chemicals containing phosphorus, such as malathion and parathion, intended to control insects. These compunds are short-lived and, therefore, do not normally contaminate the environment. However, some organophosphates, such as parathion, are extremely toxic when initially applied and exposure to them can interfere with the normal processes of the nervous system, causing convulsions and eventually death. Malathion, on the other hand, is low in toxicity and relatively safe for humans and animals; it is a common ingredient in household insecticide products.

Orifice. An opening with a sharp-edged uniform closed perimeter.

Orifice Scrubbers. Devices for the removal of particulates from gas streams in which the flow of air through a restricted passage partially filled with water causes the dispersion of the water and consequent wetting and collection of the particulates.

Orographic Lifting. The rise of airmass weather and pollutants caused by mountains.

Orsat. An apparatus used for analyzing flue gases volumetrically.

Os. See osmium.

Osmium (Os). A bluish-white hard crystalline metal of the platinum family of elements with an atomic weight of 191.5 and atomic number of 76. It is used in lamp filaments and as a hardening alloy.

OSW. Office of Saline Water, U.S. Department of the Interior.

OTA. Office of Technology Assessment (U.S. Congress).

OTEC. Ocean thermal electric conversion.

Outfall. The mouth of a sewer, drain or conduit where an effluent is discharged into the receiving waters.

OVCST. Weather abbreviation for "overcast."

Overburden. Material of any nature, consolidated or unconsolidated, that overlies a deposit of useful material, ores, or coal, especially those deposits that are mined from the surface by open cuts.

Overfire. Air for combustion admitted into the furnace at a point above the fuel bed.

Overfire Air. Air forced into the top of an incinerator to fan the flame.

Oxbow. A stretch of river channel which has extreme curvature, the channel forming one or more loops or circles with a narrow section of land separating two points on the stream at the neck or location where curvature commences.

Oxidant. Any oxygen containing substance that reacts chemically in the air to produce new substances. Oxidants are the primary contributors to photochemical smog.

Oxidation. A chemical reaction in which oxygen unites or combines with other elements. Organic matter is oxidized by the action of aerobic bacteria; thus oxidation is used in waste water treatment to break down organic wastes.

Oxidation Pond. A man-made lake or pond in which organic wastes are reduced by bacterial action. Often oxygen is bubbled through the pond to speed the process.

Oxides of Sulfur (SO_x). A collective term for the various oxides of sulfur e.g. SO_2, SO_3, SO_4.

Oxygen (O_2). A gaseous element occurring naturally in the atmosphere and constituting about 1/5 of the atmosphere. The atomic number is 8 and atomic weight is 16.00. It is very reactive in compounds and is essential to life.

Oxygen Demand Index. A determination used to approximate the 5-day BOD of wastewater and sewage samples.

Oxygen Lancing. In steel making, a procedure in which oxygen is injected into the bath of molten metal through a water cooled lance. The oxygen oxidizes carbon, silicon, manganese, and some iron to exothermic reactions. The procedure materially shortens the time needed to tap the furnace.

Ozone (O_3). A pungent, colorless, toxic gas. Ozone is one component of photochemical smog and is considered a major air pollutant. A layer of ozone 12–15 miles above the earth's surface, however, acts as a screen for the sun's ultra-violet rays.

P

P. See phosphorous.

Pa. Arterial oxygen pressure; also pascal; also protoactinuium.

PAB. Para-aminobiphenyl; a carcinogenic compound formerly used in synthetic rubber process; strongly suspected to cause bladder cancer.

Package Plant. A prefabricated or prebuilt waste water treatment plant.

Packed Column (Packed Scrubber or Packed Tower). A vertical column used for distillation, absorption, and extraction, containing packing; e.g., Raschig rings, Berl saddles, or crushed rock, which provide a large contacting surface area between phases. Normally gas flow is countercurrent to liquid flow.

Packed Tower. An air pollution control device in which polluted air is forced upward through a tower packed with crushed rock or wood chips while a liquid is sprayed downward on the packing material. The pollutants in the air stream either dissolve or chemically react with the liquid.

Palladium (Pd). A steel white metal of the platinum family of elements with an atomic weight of 106.7 and an atomic number of 46. It is used in balances and non-magnetic watches.

PAN. See peroxyacetyl nitrate.

PAO_2. Alveolar partial pressure of oxygen.

Para-Aminobiphenyl (PAB). See PAB.

Parabolic Collector. A device which uses a reflective surface shaped in the form of a parabola to concentrate solar energy on a small collector surface equipped with a means by which energy may be absorbed and carried away.

Parent. A radioactive nucleus which disintegrates to form a radioactive product or daughter. Also called precursor.

Parshall Flume. A device named after its designer for measuring the flow of a liquid in an open conduit.

Particle Concentration. Concentration expressed in terms of number of particles per unit volume of air or other gas.

Particle Size. An expression for the size of liquid or solid particles expressed as the average or equivalent diameter.

Particle Size Distribution. The relative percentage of weight or number of each of the different size fractions of particulate matter.

Particulate Loading. The introduction of particulates into the ambient air.

Particulate Matter. See particulates.

Particulates. Finely divided solid or liquid particles in the air or in an emission. Particulates include dust, smoke, fumes, mist, spray and fog.

Pathogenic. Causing or capable of causing disease.

Pb. See lead.

PBB. Polybrominated Biphenyls.

PCB. See polychlorinated biphenyl.

PCE. See pyrometric cone equivalent.

Pd. See Palladium.

Peat. Partially decomposed organic material.

Pellicular Water. Suspended water adhering by molecular attraction to wall surfaces in addition to hygroscopic adherence. It can be moved by root absorption or evaporation but not by gravity.

Peneplain. A relatively flat featureless plain resulting from erosion by streams of overlying formations.

Percolation. Downward flow or infiltration of water through the pores or spaces of a rock or soil.

Percolator. A device used in rendering plants for the separation of dry proteinaceous crackling from the clear moisture free tallow. They are generally perforated pans which allow the tallow to drain away from the cracklings.

Performance Test. Measurements of emissions used for the purpose of determining compliance with a standard of performance.

Permeability. The capacity of a porous medium to conduct or transmit fluids.

Peroxyacetyl Nitrate (PAN). A pollutant created by the action of sunlight on hydrocarbons and nitrogen oxides in the air. PAN is an integral part of photochemical smog.

Persistent Pesticides. Pesticides that will be present in the environment for longer than one growing season or one year after application.

PERT. Program evaluation and review technique – a system of planning and review of projects for programs using schedules, milestones, time sequences elements and critical paths in functional flow diagrams.

Pesticide. An agent used to control pests. This includes insecticides for use against harmful insects; herbicides for weed control; fungicides for control of plant diseases; rodenticides for killing rats, mice, etc.; and germicides used in disinfectant products, algaecides, slimicides, etc. Some pesticides can contaminate water, air or soil and accumulate in man, animals and the environment, particularly if they are misused. Certain of these chemicals have been shown to interfere with the reproductive processes of predatory birds and possibly other animals.

Pesticide Tolerance. A scientifically and legally established limit for the amount of chemical residue that can be permitted to remain in or on a harvested food or feed crop as a result of the application of a chemical for pest control purposes. Such tolerances or safety levels, established federally by EPA, are set well below the point at which residues might be harmful to consumers.

Petrochemical Industry. A branch of the petroleum industry in which refined crude oil is manufactured into various chemicals.

Petroleum Coke. See coke, petroleum.

pH. A measure of the acidity or alkalinity of a material, liquid or solid, pH is represented on a scale of 0 to 14 with 7 representing a neutral state, 0 representing the most acid and 14, the most alkaline.

PHA. Phytohemagluttinin.

Phenols. A group of organic compounds that in very low concentrations produce a taste and odor problem in water. In higher concentrations, they are toxic to aquatic life. Phenols are byproducts of petroleum refining, tanning and textile, dye and resin manufacture.

Phosphates. Phosphorous containing compounds, which are contributed to wastewaters by detergents, human excrement, and other sources. As a plant nutrient, phosphorous is necessary for the growth of algae and aquatic weeds in fresh water lakes and rivers. Death of masses of these plants creates an excessive oxygen demand which in turn leads to death of fish.

Phosphorous. A colorless waxy solid element which occurs naturally only in allotropic forms. It is poisonous but has uses because of its igniting characteristic in air. Atomic weight is 31.02 and atomic number is 15. Phosphorus is an element that while essential to life, contributes to the eutrophication of lakes and other bodies of water.

Photic Zone. Depth zone in a lake in which enough light is available for photosynthesis.

Photochemical Conversion. A chemical reaction which converts light to useable energy.

Photochemical Oxidants. Secondary pollutants formed by the action of sunlight on the oxides of nitrogen and hydrocarbons in the air; they are the primary contributors to photochemical smog.

Photochemical Process. The chemical changes brought about by the radiant energy of the sun acting upon various polluting substances. The products are known as photochemical smog.

Photochemical Reaction. A chemical reaction which involves either the absorption or emission of radiation in the form of light energy.

Photochemical Smog. Air pollution associated with oxidants rather than with sulfur oxides, particulates, etc. Produces necrosis, chlorosis and growth alterations in plants and is an eye and respiratory irritant in humans.

Photon. See quantum.

Photosynthesis. The formation of complex organic compounds from simple inorganic under the stimulus of light and the presence of chlorophyll.

Photovoltaic Conversion. The conversion of light from the sun into electricity by solid state devices.

Phreatic Water. Ground water in a zone of saturation having a water table. Non-artesian ground water.

PHS. Public Health Service.

Physical-Chemical Treatment. Unit processes involving physical and chemical as opposed to biological means to remove organic and inorganic contamination. Examples: Clarification, filtration, ion exchange, carbon absorption and reverse osmosis.

Phytoplankton. The plant portion of plankton.

Phytotoxic. Injurious to plants.

Piezometer. An instrument for measuring the pressure in a conduit, tank, or ground waters.

Piezometric Head. The distance above or below the land surface to the elevation to which the water in an artesian aquifer or ground-water lens or basin would rise at a given point if free to do so-artesian head.

Piezometric Pressure. Artesian pressure upwelling from aquifer or water bearing formation.

Pig. A container usually made of lead used to ship or store radioactive materials.

Pile. A nuclear reactor.

PL. Public Law.

Plankton. The floating or weakly swimming plant and animal life in a body of water, often microscopic in size.

Plant Section. A part of a plant consisting of one or more unit operations including auxiliary equipment which provides the complete processing of input (raw or feed-stock) materials to produce a marketable product.

Platinum (pt). A tin-white lustrous metal element which is malleable and ductile. It has an atomic weight of 195.23 and an atomic number of 78. It is used naturally and as an alloy or plating.

Playa. A lake bed in the lowest part of an enclosed valley in an arid area whose drainage of infrequent precipitated water coverage is inward.

Plume. The visible emission from a flue or chimney.

Plume Rise. The rise of a plume center line or center of mass above its point of origin due to initial vertical momentum or buoyancy, or both.

Plutonium (Pu). Radioactive element occuring in earth's surface in pitchblende. Atomic weight (most stable isotope) 242,244; atomic number 94.

PM. Particulate Matter.

PNA. Polynuclear aromatics.

PNL. Perceived Noise Level.

PNS. Permanent Network Stations.

Pneumatic Ash Handling. A system of pipes and cyclone separators that convey fly ash or floor dust to a bin via an air stream.

Po. See polonium.

POD. Point of Discharge.

Point Source. In air pollution, a stationary source of a large individual emission, generally of an industrial nature. This is a general definition; point source is legally and precisely defined in Federal regulations. See area source.

Pollen. A fine dust produced by plants; a natural or background air pollutant.

Pollutant. Any introduced gas, liquid or solid that makes a resource unfit for a specific purpose.

Polluted Water. Water that contains sewage industrial waste, or other harmful, hazardous, unhealthy or objectionable substances.

Pollution. The presence of matter or energy whose nature, location or quantity produces undesired environmental effects.

Polonium (Po). A radioactive element with an atomic weight of approximately 210 and an atomic number of 84. It is also called Radium F.

Polychlorinated Biphenyls (PCBs) A group of organic compounds used in the manufacture of plastics. In the environment PCB's exhibit many of the same characteristics as DDT and may be confused with that pesticide. PCB's are highly toxic to aquatic life, they persist in the environment for a long time, and they are biologically accumulative.

Polycyclic Molecule. A molecule containing two or more fused rings (as in anthracine).

Polyelectrolytes. Synthetic chemicals used to speed flocculation of solids in sewage.

Polymerization. A reaction combining two or more molecules to form a single molecule having the same elements in the same proportions as in the original molecules. The unit of light olefins to form hydrocarbons of higher molecular weight. The process may be thermal or catalytic.

Polynuclear Aromatic Hydrocarbons. Compounds consisting of two or more aromatic rings which share a pair of carbon atoms. The simplest and most important is naphthalene ($C_{10}H_8$; also polycyclic).

POM. Polycyclic organic matter.

Porosity. Ratio of the volume in any porous material that is not filled with solid matter to the total volume.

Positron. A subatomic particle equal in mass and weight to the electron and having an equal but opposite charge. Positrons are emitted by some artifically radioactive isotopes.

Potable Water. Water suitable for drinking or cooking purposes from both health and aesthetic considerations.

Potassium (k). A soft bright silvery metal of the alkali group with the atomic weight of 39.096 and atomic number 19. Potassium is never found free because it reacts with moisture.

Potential Energy. Energy inherent in a mass because of its position with reference to other masses; e.g. a rock at the edge of a precipice has potential energy.

Potential Temperature. The temperature that a gas would obtain if it were adiabatically compressed to some standard pressure, usually 1000 mb in meteorological literature.

POTW. Publicly Owned Treatment Work – EPA.

PPB or ppb. Parts per billion.

PPM or ppm. Parts per million. The unit commonly used to represent the degree of pollutant concentration where the concentrations are small. Larger concentrations are given in percentages. Thus BOD is represented in ppm while suspended solids in water are expressed in percentages. In air, ppm is usually a volume/volume ratio; in water, a weight/volume ratio.

Pr. See praseodymium.

Praseodymium (Pr). A metallic rare earth element whose atomic weight is 140.92 and atomic number is 59. It forms green salts.

Preaeration. A preparatory treatment of sewage comprising aeration to remove gases, add oxygen, or promote flotation of grease and aid coagulation.

Precipitate. A solid that separates from a solution because of some chemical or physical change or the formation of such a solid.

Precipitators. In pollution control work, any of a number of air pollution control devices usually using mechanical/electrical means to collect particulates from an emission.

Precleaners. Collectors of limited efficiency used ahead of the final cleaner. If the gas contains an appreciable amount of hard, coarse particles, a precleaner can materially reduce erosive wear of the more efficient final collector.

Precursors. Gaseous air pollutants which react with other substances in the atmosphere to produce different pollutants; e.g., photochemical reactions of NO and NO_2 with the oxygen of the air which produce ozone.

Pretreatment. In waste water treatment, any process used to reduce pollution load before the waste water is introduced into a main sewer system or delivered to a treatment plant for substantial reduction of the pollution load.

Prilling. A combination spray drying and crystallization technique used in the production of ammonium nitrate. A hot ammonium nitrate solution is sprayed in the top of a tower, and air is blown in at the bottom. The liquid is converted into spherical pellets.

Primary Air. In incineration, air which is introduced with the refuse into the primary chamber.

Primary Emission. Pollutants emitted directly into the air from identifiable sources.

Primary Standard. The national primary ambient air quality standard which defines levels of air quality which are necessary to protect public health.

Primary Treatment. The first stage in waste water treatment in which substantially all floating or settleable solids are mechanically removed by screening and sedimentation.

PRM. Program Requirement Memorandum (EPA).

Process. Any action, operation or treatment, and all methods and forms of manufacturing, fabricating or handling.

Process Emission Source. One or more units of processing equipment which may be operated independently of other parts of the operations at any given manufacturing or processing facility and which may emit smoke, particulate matter, gaseous matter or other air contaminant. Also, where it is common practice to group more than one unit or like or similar processing equipment together and to apply a single or combined unit of air pollution control equipment to the entire group.

Process Weight. The total weight of all materials, including fuels, introduced into a manufacturing process. The process weight is used to calculate the allowable rate of emission of pollutant matter from the process.

Processing. Any method, system or other treatment designed to change the physical form or chemical content of solid waste.

Profundal Zone. The deep region of a lake which lies below the light controlled limit of plant growth.

Properties. Characteristics by which a substance may be identified. Physical properties describe its state of matter, color, odor and density; chemical properties describe its behavior in reaction with other materials.

Protoactinium (Pa). The first of the radioactive elements in the actinium series with an atomic weight of 231 and an atomic number of 91. It is also known as EKATANTALUM and as URANIUM X_2.

Proton. See atom.

PSC. Primary Standard Condition (EPA).

PSD. Prevention of Significant Deterioration.

PSI. Pollution Standard Index; national ambient air standard index promulgated by EPA for use in air quality reporting in local news media.

psia. Pounds per square inch absolute.

psig. Pounds per square inch gage.

PSM. Point Source Monitoring.

Psychrometer. Instrumentation consisting of two similar thermometers with the bulb of one immersed in water so that the evaporative cooling loss makes it register lower temperatures. The difference between the readings is a measure of the dryness of the atmosphere.

Pt. See platinum.

Pu. Plutonium.

PUD. Planned Unit Development.

Pug Mill. A machine for mixing water and clay which consists of a long horizontal barrel within which is a long longitudinal shaft fitted with knives which slice through the clay, mixing it with water which is added by sprayers from the top. The knives are canted to give some screw action, forcing the clay along the barrel and out one end.

Pulmonary Emphysema. An anatomic change in the lungs characterized by a breakdown of the walls of the alveoli, which can become enlarged, lose their resilience, and disintegrate.

Pulmonary Function. In medicine, a term used in describing the adequacy of the lung's performance.

Pulverization. The crushing or grinding of material into small pieces.

Pump, Reciprocating. A positive-displacement type of pump consisting of a plunger or a piston moving back and forth within a cylinder. With each stroke of the plunger or piston, a definite volume of liquid is pushed out through the discharge valves.

Pumping Station. A station at which sewage is pumped to a higher level. In most sewer systems pumping is unnecessary; waste water flows by gravity to the treatment plant.

Putrefaction. Biological decomposition of organic matter with the production of ill-smelling products associated with anaerobic conditions.

Putrescible. Capable of being decomposed by microorganisms with sufficient rapidity to cause nuisances from odors, gases, etc. For example, kitchen wastes or dead animals.

Pychnometer. A device used for the measurement of densities of liquids.

Pyrheliometer. An instrument for measuring the sun's radiant energy as received on earth.

Pyritic Sulfur. Sulfur combined with iron, found in coal.

Pyrolysis. The chemical decomposition of a material by heat in the absence of oxygen.

Pyrometer. An instrument for measuring temperatures beyond the range of thermometers.

Pyrometric Cone Equivalent (PCE). An index to the refractoriness of a material; it is obtained by a test that provides the number of a standard pyrometric cone that is closest in its bending behavior to that of a pyrometric cone made of the material when both are heated in accordance with the ASTM Standard Method of Test for Pyrometric Cone Equivalent of Refractory Materials.

Q

Q. Cardiac output.

QA. Quality Assurance.

QCP. Quiet Communities Program – a noise control program sponsored by EPA.

QRS. A particular part of the wave of electrocardiogram.

Quads. Quadrillions of BTU's of energy.

Quantum. The smallest indivisible quantity of radiant energy; a photon.

Quartz. A crystal or crystalline mineral composed of silicon oxide.

Quench Tank. A water-filled tank used to cool incinerator residues.

Quercetin. A yellow crystalline pigment occurring usually in the form of glycosides in various plants.

Quicklime. A calcined material the major part of which is calcium oxide or calcium oxide in association with magnesium oxide capable of slaking in water.

Quicksand. Sand which has lost its grain to grain contact by upwelling water and has no load bearing capacity. It is a condition, not a type of material.

Quicksilver. See mercury.

Quinacrine. An antimalarial drug derived from acridine and used especially as dihydrochloride.

R

°**R**. Degrees Rankine.

RA. Regional Agency, EPA.

Ra. See Radium.

Rack Bar. A screen composed of parallel bars, either vertical or inclined, placed in a waterway to catch floating debris, and from which the screenings may be raked.

RACM. Reasonably Available Control Measures, EPA.

RACT. Reasonably Available Control Technology (EPA).

rad. A unit of measurement of any kind of radiation absorbed by man.

RADHAZ. Radiation hazards.

Radiant Energy. The transmission of energy through space or a material medium in the form of electromagnetic waves.

Radiation. The emission of fast atomic particles or rays by the nucleus of an atom. Some elements are naturally radioactive while others become radioactive after bombardment with nuetrons or other particles. The three major forms of radiation are alpha, beta and gamma.

Radiation Standards. Regulations that include exposure standards, permissible concentrations and regulations for transportation.

Radical. A group of atoms of different elements acting as a single unit in a chemical reaction; normally incapable of separate existence. A radical may be negatively charged, positively charged, or without a charge.

Radioactivity. The nuclear energy released when the nucleus of an atom disintegrates.

Radiobiology. The study of the principles, mechanisms and effects of radiation on living matter.

Radioecology. The study of the effects of radiation on species of plants and animals in natural communities.

Radioisotopes. Radioactive isotopes. Radioisotopes such as Cobalt-60 are used in the treatment of disease.

Radiosonde. A miniature radio transmitter carried aloft by an unmanned balloon to transmit humidity, temperature, and pressure data.

Radium (Ra). A radioactive brilliant white metallic element whose atomic weight is 226.05 and atomic number is 88. It is used for its luminosity and in medicine for treatment of cancer.

Radium F. See polonium.

Radon (Rn). A heavy shining gaseous element which emanates from radium. Its atomic weight is 222.0 and its atomic number is 86.

RAM. Real Time Air Quality Model, EPA.

Rasp. A device used to grate solid waste into a more manageable material ridding it of much of its odor.

Ravine. An elongated depression, larger than a gulley and smaller than a valley.

R_{aw}. Airway resistance.

Raw Sewage. Untreated domestic or commercial waste water.

Rawinsonde. A radiosonde tracked by a radio-direction finder to determine wind velocities aloft.

Rb. See Rubidium.

RBC. Red blood cell.

RBW. Receiving Body of Water.

RC & RA. Federal Resource Conservation and Recovery Act of 1976.

RDF. Refuse derived fuel.

Re. See Rhenium.

Reactor. The apparatus in which nuclear fission takes place.

Reasonably Available Control Technology (RACT). An existing stationary source – the extent of emission control technology determined by case-by-case analyses to be economically and technologically reasonable requirements for emission control.

Receiving Waters. Rivers, lakes, oceans or other bodies that receive treated or untreated waste waters.

Receptor. A location at which ambient air quality is measured or estimated.

Reciprocating Engine. An engine in which the up-and-down motion of a piston is transformed into the circular motion of the crankshaft.

Recoverable Resources. Materials that still have useful or chemical properties after serving a specific purpose and can, therefore, be reused or recycled for the same or other purposes.

Recovery Boiler. In wood pulping, a combustion unit designed to recover the spent chemicals from the cooking liquor and to produce steam for pulping and recovery operations.

Recycling. The process by which waste materials are transformed into new products in such a manner that the original products may lose their identity.

Red Tide. A proliferation or bloom of a certain type of plankton with red-to-orange coloration, that often causes massive fish kills. Though they are a natural phenomenon, blooms are believed to be stimulated by phosphorus and other nutrients discharged into waterways by man.

Reductant. Organic matter in streams and sewage which is stabilized under anaerobic conditions.

Reduction. The addition of hydrogen or the abstraction of oxygen from a substance; the extraction of any metal from its ore.

Re-Entrainment. The resuspension in the atmosphere of particles from streets, rooftops, etc., by wind, passing vehicles, or other such forces.

Reference Conditions. EPA requires that all measurements of air quality be corrected to a reference pressure of 760 millimeters of HG (1,013.2) millibars.

Refinery Gas. Any form of mixture of still gas gathered in a refinery from the various stills.

Refining. In metallurgy, the removal of impurities necessary to produce an ingot or alloy of desired specification. In petroleum, the process of separating, combining, or rearranging petroleum oil constituents to produce salable products.

Reflectivity. The ratio of light reflected from a surface to the light falling on the surface. The reflectivity plus transmissivity and absorptivity equals one, since upon all surfaces the incident sunlight is transmitted, reflected or absorbed.

Reforming. The thermal or catalytic conversion of naphtha into more volatile products of higher octane number. It represents the total effect of numerous reactions, such as cracking, polymerization, dehydrogenation, and isomerization, taking place simultaneously.

Refractory. A ceramic material of a very high melting point with properties that make it suitable for such uses as furnace and kiln linings.

Refuse. See solid waste.

Refuse Reclamation. The process of converting solid waste to saleable products. For example; the composting of organic solid waste yields a saleable soil conditioner.

REGS. Regulations.

Reinjection. Reintroduction of fly ash into a furnace to burn out all the combustibles.

Relative Humidity. The ratio of the actual quantity of water vapor present in a small volume to the quantity required for saturation at the same temperature – the percentage of saturation.

rem. A measurement of radiation dose to the internal tissues of man.

rep. A unit of measurement of any kind of radiation absorbed by man.

Rerun Oil. Oil which has been redistilled.

Reservoir. A pond, lake, tank or basin, natural or man-made, used for the storage, regulation and control of water.

Residual. Heavy oil left in the still after gasoline and other distillates have been distilled off, or residue from the crude oil after distilling off all but the heaviest components.

Residual Chlorine. The quantity of chlorine, in excess of the chlorine demand remaining in water, sewage, or effluents, after a selected contact period of time, expressed in parts per million. The difference between the chlorine dose and the chlorine demand.

Resistivity. The property of a body whereby it opposes and limits the passage of electricity through it. Resistivity of dust is an important factor in the performance of electrostatic precipitators. If the resistivity of the collected dust is higher than about 2×10^{10} ohm-cm, excessive arcing or reverse corona can occur, thereby limiting precipitator performance.

Resource Recovery. The process of obtaining materials or energy, particularly from solid waste.

Retrofitting. To change the design of an existing structure, tool or machine.

Reverberation. The persistence of sound in an enclosed space after the sound source has stopped.

Reverberatory Furnace A furnace with a shallow hearth; having a roof that deflects the flame and radiates heat toward the surface of the charge. Firing may be with coal, pulverized coal, oil, or gas.

Reverse Osmosis. The extraction of a dilute solution from a more concentrated or saturated solution by the application of over pressures to exude the dilute through a permeable membrane. The process has been used successfully in wastewater treatment plants.

RFP. Request for proposal; also, Reasonable Further Progress, EPA.

Rh. See Rhodium.

Rhenium (Re). A rare metallic element which dissolves in nitric acid and sulfuric acid. It has an atomic weight of 186.31 and an atomic number of 75.

Rhodium (Rh). A silver white metallic element of the platinum family whose atomic weight is 102.91 and atomic number is 45. It is used in alloys for hardening surfaces and for appearance.

Rill. A small brook.

Ringelmann Chart. A series of illustrations ranging from light grey to black used to measure the opacity of smoke emitted from stacks and other sources. The shades of grey similate various smoke densities and are assigned numbers ranging from one to five. Ringelmann No. 1 is equivalent to 20 per cent dense; No. 5 is 100 per cent dense. Ringelmann charts are used in the setting and enforcement of emission standards.

Ringelmann Number. See Ringelmann Chart.

Riparian Rights. Rights of a land owner to the water on or bordering his property. Including the right to prevent diversion or misuse of upstream water.

River Basin. The total area drained by a river and its tributaries.

RL. Regional Limitations (EPA).

RMA. Rubber Manufacturers Association.

RMS. Root-mean-square.

Rn. See Radon.

RNA. Ribonucleic acid.

RO. Regional Office (RO).

Roast. To heat to a point somewhat short of fusing, with access to air, so as to expel volatile matter or effect oxidation. In copper metallurgy, applied specifically to the final heating which causes self-reduction to occur by the reaction between the sulfide and the oxide.

Roaster. A contrivance for roasting, or a furnace for drying salt cake. A reverberatory furnace or a muffle used in roasting ore.

Roasting. Heating an ore to effect some chemical change that will facilitate smelting. The heating of solids, frequently to promote a reaction with a gaseous constituent in the furnace atmosphere.

Roasting Furnace. A furnace in which finely ground ores and concentrates are roasted to eliminate sulfur; heat is provided by the burning sulfur.

Rodenticide. A chemical or agent used to destroy or prevent damage by rats or other rodent pests. See pesticide.

Roentgen. The original unit of measurement of X or gamma radiation. See rad and rep. Also written, r, or r/unit.

Roll Back. A simple air quality model that assumes that if emissions from each source affecting a given receptor are decreased by the same percentage, ambient air quality concentrations decrease proportionately.

Rough Fish. Those fish species considered to be of poor fighting quality when taken on tackle or of poor eating quality; for example, gar, suckers, etc. Most rough fish are more tolerant of widely changing environmental conditions than are game fish.

RPAR. Rebuttable presumption against registration (EPA).

RPM. Revolutions per minute.

RPS. Revolutions per sec.

RTCM. Reasonable Transportation Control Measures, EPA.

RTP. Research Triangle Park near Durham, N.C. Site of EPA research center and Air Pollution Institute.

Ru. See Ruthenium.

Rubbish. A general term for solid waste – excluding foodwaste and ashes – taken from residence, commercial establishments and institutions.

Rubidium (Rb). A soft white rare metal element of the potassium group. Atomic weight is 85.44 and atomic number 37.

Run of Mine Coal. Unscreened bituminous coal as it comes from the mine.

Runoff. The portion of rainfall, melted snow or irrigation water that flows across ground surface and eventually is returned to streams. Runoff can pick up pollutants from the air or the land and carry them to the receiving waters.

Ruthenium (Ru). A hard brittle grey metal of the platinum family whose atomic weight is 101.7 and atomic number is 44.

RW. Weather abbreviation for "rain showers."

S

S. See sulfur.

s. Second.

S & A. Surveillance and Analysis (element of EPA Regional Office).

SAB. Science Advisory Board, (EPA).

Salamander. A small portable incinerator, or a small portable heater burning coke or oil.

Salinity. The degree of salt in water.

Salt Water Intrusion. The invasion of salt water into a body of fresh water, occurring in either surface or groundwater bodies. When this invasion is caused by oceanic waters, it is called sea water intrusion.

Salvage. The utilization of waste materials.

Samarium (Sm). A rare earth element whose atomic weight is 150.43 and atomic number is 62.

SAMWG. Standing Air Monitoring Work Group (EPA).

Sand Filter. An interim or temporary waste water treatment facility used where the water table is too high to accommodate a septic tank and central treatment facilities are not available. Sand and/or coarse aggregate are placed in an enclosed area above grade and soak up wastes. System must be scarified on a regular basis and medium must be replaced periodically.

Sanitary Landfill. A site for solid waste disposal using sanitary land-filling techniques.

Sanitary Landfilling. An engineered method of solid waste disposal on land in a manner that protects the environment; waste is spread in thin layers, compacted to the smallest practical volume and covered with soil at the end of each working day.

Sanitary Sewers. Sewers that carry only domestic or commercial sewage. Storm water runoff is carried in a separate system. See sewer.

Sanitation. The control of all the factors in man's physical environment that exercise or can exercise a deleterious effect on his physical development, health and survival.

Saprobic. An environment rich in organic matter and relatively low in oxygen.

SAROAD. Storage and Retrieval of Aerometric Data (EPA ambient air quality data system).

Saturated. Said of an organic compound that cannot incorporate additional hydrogen atoms into its structure; such compounds are not as reactive as unsaturated ones. Also, the state of a substance which has absorbed all of the moisture that it can.

Saturation. That point in a process at which a substance can no longer absorb additional moisture – soaked.

Sb. See Antimony.

SBA. Small Business Administration.

Sc. See Scandium.

Scandium (Sc). A rare earth metal element whose atomic weight is 45.10 and atomic number is 21. It forms colorless salts upon oxidation.

scf. Standard cubic feet.

scfh. Cubic feet per hour at standard conditions.

SCFM. See Standard cubic feet per minute.

Schmutzdecke. A "dirty skin" or layer of flocculent material that forms on the surface of a sand filter.

Scrap. Discarded or rejected materials that result from manufacturing or fabricating operations and are suitable for reprocessing.

Screening. The removal of relatively coarse floating and suspended solids by straining through racks or screens.

Screening Procedure. In air quality modelling, a relatively simple analysis technique to determine if a given source is likely to pose a threat to air quality.

Scrubber. An air pollution control device that uses a liquid spray to remove pollutants from a gas stream by absorption or chemical reaction. Scrubbers also reduce the temperature of the emission.

SCS. Soil Conservation Service (USDA); also Supplemental Control System, also Secondary Control System.

SDA. Soap and Detergent Association.

Se. See Selenium.

sec. Second.

Secondary Air. Air introduced into a combustion chamber beyond the point of fuel and primary air introduction for the purpose of achieving more complete oxidation.

Secondary Standard. The national secondary ambient air quality which defines levels of air quality judged necessary to protect the public welfare from any known or anticipated adverse effects of a pollutant.

Secondary Treatment. Waste water treatment, beyond the primary stage, in which bacteria consume the organic parts of the wastes. This biochemical action is accomplished by use of trickling filters or the activated sludge process. Effective secondary treatment removes virtually all floating and settleable solids and approximately 90 per cent of both BOD_5 and suspended solids. Customarily, disinfection by chlorination is the final stage of the secondary treatment process.

Sedimentation. In waste water treatment, the settling out of solids by gravity.

Sedimentation Tanks. In waste water treatment, tanks where the solids are allowed to settle or to float as scum. Scum is skimmed off; settled solids are pumped to incinerators, digesters, filters or other means of disposal.

Seeding Sludge. The introduction or inoculation of undigested sewage solids with sludge that has undergone decomposition, for the purpose of introducing favorable organisms accelerating the formation of the biomass.

Seepage. Water that flows through the soil.

SEER. Seasonal Energy Efficiency Ratio (DOE).

Selective Coating. An optical coating for collectors that normally has a high absorptivity for incident sunlight and low emissivity for long wave radiation.

Selective Herbicide. A pesticide intended to kill only certain types of plants, especially broad-leafed weeds, and not harm other plants such as farm crops or lawn grasses. The leading herbicide in the United States is 2,4-D. A related but stronger chemical used mostly for brush control on range, pasture, and forest lands and on utility or highway rights-of-way is 2,4,5-T. Uses of the latter chemical have been somewhat restricted because of laboratory evidence that a dioxin contaminant in 2,4,5-T can cause birth defects in test animals.

Selenium (Se). A gray, crystalline semi-metallic element of the sulfur group whose atomic weight is 78.96 and atomic number is 34. It is used in glass and ceramics.

SELS. Weather abbreviation for "severe local storm."

Senescence. The process of growing old. Sometimes used to refer to lakes nearing extinction.

Septic Tank. An underground tank used for the decomposition of domestic wastes. Bacteria in the wastes decompose the organic matter, and the sludge settles to the bottom. The effluent flows through drains into the ground. Sludge is pumped out at regular intervals.

Septicization. A term applied to anaerobic decomposition, whereby intensive growths of bacteria with the enzymes secreted by them liquefy and gasify solid organic matters.

SERI. Solar Energy Research Institute – Colorado.

SERS. Seaborne environmental reporting system.

Set. The direction that a current of water is going toward.

Settleable Particulate. Particulate matter which is emitted into the atmosphere such that it may deposit onto horizontal surfaces due to gravitational settling.

Settleable Solids. Bits of debris and fine matter heavy enough to settle out of waste water.

Settling Chamber. In air pollution control, a low-cost device used to reduce the velocity of flue gases usually by means of baffles, promoting the settling of fly ash.

Settling Tank. In waste water treatment, a tank or basin in which settleable solids are removed by gravity.

Sewage. The total of organic waste and waste water generated by residential and commercial establishments.

Sewage Lagoon. See lagoon.

Sewer. Any pipe or conduit used to collect and carry away sewage or stormwater runoff from the generating source to treatment plants or receiving streams. A sewer that conveys household and commercial sewage is called a sanitary sewer. If it transports runoff from rain or snow, it is called a storm sewer. Often storm water runoff and sewage are transported in the same system or combined sewers.

Sewerage. The entire system of sewage collection, treatment and disposal. Also applies to all effluent carried by sewers whether it is sanitary sewage, industrial wastes or storm water runoff.

SGOT. Serum glutamic-oxaloacetic transaminase.

SGPT. Serum glutamic – pyruvic transaminase.

Shield. A wall that protects workers from harmful radiation released by radio-active materials.

Shutdown. The cessation of the operation of a source for any purpose.

Si. See silicon.

SIC. Standard Industrial Classification (Chemicals) (EPA).

Silicon (Si.). A commonly occurring non-metallic element resembling graphite whose atomic weight is 28.06 and atomic number is 14.

Silt. Finely divided particles of soil or rock. Often carried in cloudy suspension in water and eventually deposited as sediment.

Silt Barrier. A berm, embankment, or anchored curtain to contain silt temporarily in drainoff from a construction project.

Silver (Ag). A pure white metal, known to the ancients, having a brilliant luster which excels over all others as a conductor. Atomic weight is 107.880 and atomic number is 47.

Silviculture. The development and care of forests. A silviculture fire is a controlled underbrush fire to rid a forest of inhibiting growths.

Sinking. A method of controlling oil spills that employs an agent to entrap oil droplets and sink them to the bottom of the body of water. The oil and sinking agent are eventually biologically degraded.

Sintering. A heat treatment that causes adjacent particles of material to cohere or agglomerate at a temperature below that of complete melting.

SIP. State Implementation Plan.

Skimming. The mechanical removal of oil or scum from the surface of water.

Skimming Plant. An oil refinery designed to remove and finish only the lighter constituents from the crude oil, such as gasoline and kerosene. In such a plant the portion of the crude remaining after the above products are removed is usually sold as fuel oil.

Skip Hoist, Inclined. A bucket or can operating up and down, receiving, elevating, and discharging bulk materials.

Slack Water. At any point in a coastal or estuarine area under tidal influence between flood and ebb tides diurnally a period of no tidal current exists which is known as slack water. High slack occurs at the end of flood tide and low slack occurs at the end of ebb tide.

Slag. The non-metallic top layer which separates from the metallic products in smelting of ores.

SLAMS. State and Local Ambient Monitoring Stations.

Sleek. Thin oily film usually present on surface of water into which sewage or oily waste has been discharged.

Slick. See Sleek.

SLF. Sanitary landfill.

Slop or Slop Oil. A term rather loosely used to denote odds and ends of oil produced at various places in a plant, which must be rerun or further processed in order to get in suitable condition for use. When good for nothing else, such oil usually goes into pressure-still charging stock, or to coke stills.

Sloughing. The phenomenon associated with trickling filters and contact aerators whereby slime and solids accumulated in the media are discharged with the effluent.

Sludge. The construction of solids removed from sewage during waste water treatment. Sludge disposal is then handled by incineration, dumping or burial.

Sluice. A conduit for carrying fluids at high velocities.

Slump. A type of landslide or downward slipping of a mass of unconsolidated material usually with a backward rotation on a horizontal axis parallel to the slope.

Sm. See Samarium.

Smelt. In wood pulping, the molten chemicals from the kraft recovery furnace consisting mostly of sodium sulfide and sodium carbonate.

Smelting. Any metallurgical operation in which metal is separated by fusion from impurities with which it may be chemically combined or physically mixed, such as in ores.

Smog. Generally used as an equivalent of air pollution, particularly associated with oxidants.

Smoke. Solid particles generated as a result of the incomplete combustion of materials containing carbon.

Smoke Candles. Apparatus used in collecting acid mists. Tubes or candles made from glass or plastic fibers are pressed into pads with thicknesses up to 2 inches and mounted in banks. Efficiency is much increased when the glass is treated with silicone oil to repel water, or when normally water-repellent plastic is used.

Smoke Unit. The number of "smoke units" is obtained by multiplying the smoke density in Ringelmann numbers by the time of occurrence in minutes. For the purpose of this calculation, a Ringelmann density reading is made at least once per minute during the period of observation. The sum of the Ringelmann density readings (made once per minute) during the period of observations would equal the number of smoke units.

SMSA. Standard Metropolitan Statistical Area.

SN. Suspended nitrates.

Sn. See tin.

SO_2. Sulfur dioxide.

Soda Ash. Common name for sodium carbonate.

Solifluction. The slow flowing masses of waste saturated with water from higher to lower levels.

Sodium (Na). A soft white silvery metal of the alkali group which reacts to decompose water into hydrogen oxide and oxygen. Atomic weight is 22.997 and atomic number is 11.

Soft Detergents. Biodegradable detergents.

Soft Water. Water with a low content of dissolved minerals, salts of calcium and magnesium.

Soil Conditioner. A biologically stable organic material such as humus or compost that makes soil more amenable to the passage of water and to the distribution of fertilizing material, providing a better medium for necessary soil bacteria growth.

Soiling. Visible damage to materials by deposition of air pollutants.

Solar Altitude Angle. The angle in the vertical plane between the sun and the plane of the horizon. In precise celestial navigation allowances must be made for the sun's semi-diameter, refraction, height of observer above horizon, instrument and time errors.

Solar Amplitude. The number of degrees the sun on the horizon is north or south of an EAST-WEST LINE through the observer.

Solar Azimuth. The compass direction of sun from observer measured in degrees (0°–360°) or degrees (0°–180°) from North and South with appropriate labelling.

Solar Cells. Photovoltaic cells used to convert sunlight directly into electricity.

Solar Constant. The intensity of the solar radiation on a surface normal to the sun's rays beyond the earth's atmosphere at the average earth sun distance of 92,955,888 miles. It is equal to 429 $BTU/ft^2/hr$.

Solar Furnaces. Solar concentrators which produce very high temperatures for industrial uses – Temperatures as high as 7000°F have been achieved.

Solar Irradiation. The same as solar radiation.

Solar Radiation. The energy received from the sun.

Solar Time. Time referenced to Greenwich and local longitude rather than time zones. Noon occurs when sun passes directly overhead or due north or south of observers local meridian of longitude.

Solid Waste. Useless, unwanted or discarded material with insufficient liquid content to be free flowing. Also see waste. (1) agricultural-solid waste that results from the raising and slaughtering of animals and the processing of animal products and orchard and field crops. (2) commercial – waste generated by stores, offices and other activities that do not actually turn out a product. (3) industrial – waste that results from industrial processes and manufacturing. (4) institutional waste originating from educational, health care and research facilities. (5) municipal – residential and commercial solid waste generated within a community. (6) pesticide – the residue from the manufacturing, handling or use of chemicals intended for killing plant and animal pests. (7) residential – waste that normally originates in a residential environment. Sometimes called domestic solid waste.

Solid Waste Disposal. The ultimate disposition of refuse that cannot be salvaged or recycled.

Solid Waste Field Capacity. The amount of water retained in solid waste after it has been saturated and has drained freely. Also, known as moisture-holding capacity.

Solid Waste Management. The purposeful, systematic control of the generation, storage, collection, transport, separation, processing, recycling, recovery and disposal of solid wastes.

Sonic Boom. The tremendous booming sound produced as a vehicle, usually a supersonic jet airplane, exceeds the speeds of sound, and the shock wave reaches the ground.

Soot. Agglomerations of tar-impregnated carbon particles that form when carbonaceous-material does not undergo complete combustion.

Sorption. A term including both absorption and adsorption. Sorption is basic to many processes used to remove gaseous and particulate pollutants from an emission and to clean up oil spills.

SOTDAT. Source test data system (EPA).

Sour. Gasolines, naphthas, and refined oils are said to be "sour" if they show a positive "doctor test;" i.e., if they contain hydrogen sulfide and/or mercaptans. Sourness is directly connected with odor, while a "sweet" gasoline has a good odor.

Source. Any property, real or personal, or person contributing to air pollution.

Source Sample. A sample of the emission from an air contamination source, collected for analysis from within a stack.

SO_x. See oxides of sulfur.

Spark Arrestor. A screenlike device to prevent sparks, embers and other ignited materials larger than a given size from being expelled to the atmosphere.

Specific Gravity. The ratio of the weight of a unit volume of a substance to that of an equal volume of water under standard conditions.

Specific Heat. The ratio of the amount of heat required to raise the temperature of a given weight of a given substance under standard pressure 1°F, to that required to raise the temperature of the same weight of water under standard pressure from 62°F to 63°F.

Specific Humidity. The ratio of the weight of water vapor to that of air in a given sample, in grams per kilogram.

Speiss. Metallic arsenides and antimonides smelted from cobalt and lead ores.

SPI. Society of Plastics Industry.

Spirometer. An instrument that measures the flow of air in and out of the lungs.

SPM. Special Purpose Monitoring.

Spoil. Dirt or rock that has been removed from its original location, specifically materials that have been dredged from the bottoms of waterways.

Spray Chamber. The simplest type of scrubber consisting of a chamber in which spray nozzles are placed. They are used extensively as gas coolers because they have a low collection efficiency for anything but coarse particles.

sq. ft. Square feet.

Sr. See Strontium.

SR_{aw}. Specific airway resistance.

SS. Suspended sulfates; also, suspended solids.

SSEIS. Standards Support and Environmental Impact Statement, EPA.

Stable. Possessing hydrostatic stability. A stable atmosphere has a positive potential temperature gradient.

Stable Air. An air mass that remains in the same position rather than moving in its normal horizontal and vertical directions. Stable air does not disperse pollutants and can lead to high buildups of air pollution.

Stability (Static Stability). The state of the atmosphere when it is stable relative to vertical displacements.

Stabilization. The process of converting active organic matter in sewage sludge or solid wastes into inert, harmless material.

Stabilization Ponds. See lagoon, oxidation pond.

Stack. A smokestack; a vertical pipe or flue designed to exhaust gases and suspended particulate matter; also, the term applied to vents for interior piped plumbing lines.

Stack Effect. The upward movement of hot gases in a stack due to the temperature difference between the gases and the atmosphere.

Stack Sampling. The collecting of representative samples of gaseous and particulate matter that flows through a duct or stack.

Stack Spray. A nozzle or series of nozzles installed in a stack above the breeching, used to inject wetting agents at high pressure to suppress the discharge of particulate matter from the stack.

Stagnant Atmospheric Conditions. Denotes when the atmospheric and meteorological conditions will reduce the necessary diffusion and dispersement of air pollutants in the atmosphere.

Stagnation. Lack of wind in an air mass or lack of motion in water. Both cases tend to entrap and concentrate pollutants.

Stagnation Temperature. The highest temperature a solar collector plate can reach.

Standard Conditions. For source testing, 70° F (21.1°C) and 29.92″Hg (760mm Hg); for air quality measurements, 77°F (25°C) and 29.92″Hg (760mm Hg); for chemistry, 273.1°K (0°C) and one atmosphere (760mm Hg); for petroleum refining, 60°F (15.55°C) and 14.7 psi (760mm Hg).

Standard Cubic Feet Per Minute. A unit measurement of gas flow which eliminates the possibility of confusion by incorporating a "standard" temperature and pressure in the term. The use of scfm instead of cfm or acfm is often recommended.

Startup. The commencement of operation of any source which has shutdown, stopped, or ceased operation for a period of time sufficient to cause temperature, pressure, chemical or other process or pollution control device imbalances which may result in excess emissions.

State Implementation Plan (SIP). A document prepared by each state, as required by the Federal Clean Air Act describing existing air quality conditions and setting forth a program to attain and maintain National Ambient Air Quality Standards.

Stationary Source. A pollution emitter that is fixed rather than moving as an automobile.

std. Standard.

Steam Distillation. Introduction of "open steam" into the liquid during distillation to assist in vaporizing the volatiles at a lower temperature.

Steining. The process of lining a well with bricks, stone, metal or other materials to keep the sides from caving in.

Still. A closed chamber, usually cylindrical, in which heat is applied to a substance to change it into vapor, with or without chemical decomposition. The substance, in its vapor form, is conducted to some cooling apparatus where it is condensed, liquefied, and collected in another part of the unit.

Stock. In general, any oil which is to receive further treatment before going into finished products.

Stoker. A machine for feeding coal into a furnace, and supporting it there during the period of combustion. It may also perform other functions, such as supply air, control combustion, or distill volatile matter. Modern stokers may be classified as overfeed, underfeed, and conveyor. Any mechanical device that feeds fuel uniformly onto a grate or hearth within a furnace may be termed a "stoker."

Stomata. Tiny openings on the underside of a leaf, through which a plant takes in carbon dioxide, and some polluting gases.

Stoping. In mining, any process of excavating ore which has been made accessible by shafts and drifts.

Storm Sewer. A conduit that collects and transports rain and snow runoff back to the ground water. In a separate sewerage system, storm sewers are entirely separate from those carrying domestic and commercial waste water.

Straight-Run Distillation. Continuous distillation which separates the products of petroleum in the order of their boiling points without cracking.

Stratification. Separating into layers. Also, the variation of potential temperature with height. Usually the term "stratified fluid" refers to a fluid possessing hydrostatic stability, as does the atmosphere when the potential temperature gradient is positive.

Strip Mining. A process in which rock and top soil strata overlying ore or fuel deposits are scraped away by mechanical shovels. Also known as surface mining.

Stripper. Equipment in which the lightest fractions are removed from a mixture. In a natural-gasoline plant, gasoline fractions are stripped from rich oil. In the distillation of crude petroleum, light fractions are stripped from the various products.

Strontium (Sr). A hard yellowish metal element whose atomic weight is 87.63 and atomic number is 38. Salts impart a brilliant crimson to flame and are used in pyrotechnics.

Subaqueous. Submerged under water – as a subaqueous pipe on a river bed.

Sublimation. The process in which certain solids under standard atmospheric conditions vaporize directly into a gaseous state without first going into a liquid state. e.g. mothballs.

Subsidence. The lowering of the elevation of a land area by mining, packing, underground decay, settling or drainoff of ground water.

Substitution. A chemical reaction in which one or more atoms or groups of a molecule are replaced by equivalent atoms or groups to form at least two products, especially the replacement of hydrogen in an organic compound by another element or group.

Substrate. The liquor in which activated sludge is kept in suspension.

Sulfidity. An expression of the percentage makeup of chemical kraft cooking liquor obtained by the formula

$$\frac{Na_2S}{Na_2S + NaOH} \times 100$$

where the sodium compounds are expressed as Na_2O.

Sulfur (S). A common pale yellow odorless brittle solid element with many widespread uses in compounds. Atomic weight is 32.06 and atomic number is 16.

Sulfur Dioxide (SO_2). A heavy, pungent, colorless gas formed primarily by the combustion of fossil fuels. SO_2 damages the respiratory tract as well as vegetation and materials and is considered a major air pollutant.

Sulfur Oxides. Pungent, colorless gases formed primarily by the combustion of fossil fuels; considered major air pollutants; sulfur oxides may damage the respiratory tract as well as vegetation.

Sullage. Mud and silt deposited by flowing water.

Sump. A depression or tank that serves as a drain or receptacle for liquids for salvage or disposal.

Supernatant. Floating on surface, like oil on water.

Superphosphate. Products obtained by mixing phosphate rock with either sulfuric or phosphoric acid, or both.

Surface Condensers. A condenser in which the coolant does not contact the vapors or condensate. Most are of the tube and shell type. Water flows inside the tubes and vapors condense on the shell side.

Surfactant. An agent used in detergents to cause lathering. Composed of several phosphate compounds, surfactants are a source of external enrichment thought to speed the eutrophication of our lakes,

Surge. A short term sizable increase in flow or pressure.

Surge Tank. A storage reservoir at the downstream end of a feeder pipe to absorb sudden rises of pressure and to furnish liquid quickly during a drop in pressure.

Surveillance System. A monitoring system to determine environmental quality. Surveillance systems should be established to monitor all aspects of progress toward attainment of environmental standards and to identify potential episodes of high pollutant concentrations in time to take preventive action.

Suspended Particulate. Particulate matter which will remain airborne for an appreciable period of time.

Suspended Solids (SS). Small particles of solid pollutants in sewage that contribute to turbidity and that resist separation by conventional means. The examination of suspended solids and the BOD test constitute the two main determinations for water quality performed at waste water treatment facilities.

Suspensoids. Colloidal particles which remain in suspension under all conditions and will combine and react with the liquid in which they exist to only a limited extent.

SW. Weather abbreviation for "snow showers;" southwest.

Sweetening. The process by which petroleum products are improved in odor and color by oxidizing or removing the sulfur containing any unsatured compounds.

SWIRS. Solid Wastes Information Retrieval Services (EPA).

Symbiosis. Biological action which promotes putrefaction and oxidation.

Synergism. The cooperative action of separate substances so that the total effect is greater than the sum of the effects of the substances acting independently.

Synthetic Crude. The total liquid, multi-component mixture resulting from a process involving molecular rearrangement of charge stock. Term commonly applied to the product from cracking, reforming, visbreaking, etc.

Systemic Pesticide. A pesticide chemical that is carried to other parts of a plant or animal after it is injected or taken up from the soil or body surface.

T

Ta. See tantalum.

Tail Gas. The exhaust or waste gas from a process.

Tail Oil. That portion of an oil which vaporizes near the end of the distillation; the heavy end.

Tailings. Second grade or waste material derived when raw material is screened or processed.

Tallow. The rendered fat of animals that is white and almost tasteless when pure, composed of glycerides of fatty acids containing a large proportion of palmitic acid and stearic acid, and that is used chiefly in making soap, glycerol, margarine, candles, and lubricants.

Tantalum (ta). A tenacious ductile metal element with a high point of fusion which is used in filaments and as an alloy. Atomic weight is 180.88 and atomic number is 73.

Tape Sampler. A device used in the measurement of both gases and fine particulates. It allows air sampling to be made automatically at predetermined times.

Tapping. Removing molten metal from a furnace.

Tb. See Terbium.

TCM. Transportation Control Measures.

TCP. Transportation Control Plan (USEPA).

TCS. Transportation Control Strategy (USEPA).

TDS. Total dissolved solids.

Te. See Tellurium.

Tellurium (Te). A semi-metallic element of the sulfur group whose atomic weight is 127.61 and atomic number is 62.

TEM. Texas Episodic Model.

Temperature Gradient. In meterology, usually the vertical gradient of mean temperature.

Temperature Inversion. An atmospheric layer in which temperature increases with altitude. The principal characteristic of a temperature inversion is its marked static stability, so that very little turbulent exchange can occur within it (see also inversion).

Terbium (Tb). A rare earth metal element whose atomic weight is 159.2 and atomic number is 65.

Tertiary Treatment. Waste water treatment beyond the secondary, or biological stage that includes removal of nutrients such as phosphorus and nitrogen, and a high percentage of suspended solids. Tertiary treatment, also known as advanced waste treatment, produces a high quality effluent.

Th. See Thorium.

Thallium (T_l). A highly malleable metal element resembling lead whose atomic weight is 204.39 and atomic weight is 81. A salt compound is used as a rodenticide.

Theoretical Air. The exact amount of air (stoichiometric air) required to supply the oxygen necessary for the complete combustion of a given quantity of a specific fuel or refuse.

Therm. A unit denoting 100,000 BTU's.

Thermal Pollution. Degradation of water quality by the introduction of a heated effluent. Primarily a result of the discharge of cooling waters from industrial processes, particularly from electrical power generation. Even small deviations from normal water temperatures can affect aquatic life. Thermal pollution usually can be controlled by cooling towers.

Thermal Turbulence. Air movement and mixing caused by temperature difference.

Thermophone. Instrument for determining temperatures at different depths in bodies of water.

Thermosyphon Heater. A gravity feed type solar water heater.

THM. Trihalomethane.

Thorium (Th). A radioactive metal element of the tin family found in rare minerals. It has an atomic weight of 232.12 and an atomic number of 90.

Threshold Dose. The minimum dose of a given substance necessary to produce a measurable physiological or psychological effect.

Thulium (Tm). A rare earth element of the erbium family which oxidizes to form highly colored salts. It has an atomic weight of 232.12 and an atomic number of 69.

Ti. See Titanium.

Tilt Angle. See angle of inclination.

Tin (Sn). A silver white malleable somewhat ductile metal element which is corrosion resistant and is used in plating and also in compounds as a reducing agent. Atomic weight is 118.70 and atomic number is 50.

TIP. Transportation Improvement Program.

Titanium (Ti). A lustrous white metal element of the Tin group. Only element which burns in nitrogen. Used in steel to increase tensile strength. Atomic weight is 47.09 and atomic number is 22.

TKN. Total Kjeldahl nitrogen.

Tl. See Thallium.

Tm. See Thulium.

Tolerance. The relative capability of an organism to endure an unfavorable environmental factor. The amount of a chemical considered safe on any food to be eaten by man or animals. Also see pesticide tolerance.

Topography. The configuration of a surface, including its relief and the position of its natural and man-made features.

Topped Crude Petroleum. A residual product remaining after the removal, by distillation, of an appreciable quantity of the more volatile components of crude petroleum.

Topping. The distillation of crude oil to remove light fractions only.

Total Kjeldahl Nitrogen (TKN). A water quality indicator which chemically equates to organic nitrogen plus ammonia.

Total Reduced Sulfur Compounds (TRS). Malodorous gases produced in the wood pulping industry exclusive of sulfur oxides. TRS usually includes hydrogen sulfide (H_2S), methyl merceptan (CH_3SH), dimethyl sulfide (CH_3SCH_3), and dimethyl disulfide (CH_3SSCH_3). The concentration of TRS is usually expressed as H_2S regardless of the constituent compounds.

Total Suspended Particulate (TSP). A criteria pollutant for which National Ambient Air Quality Standards have been set.

Toxicant. A substance that kills or injures an organism through its chemical or physical action or by altering its environment; for example, cyanides, phenols, pesticides or heavy metals. Especially used for insect control.

Toxicity. The quality or degree of being poisonous or harmful to plant or animal life.

TPD. Tons per day.

TPH. Tons per hour.

TPY. Tons per year.

Transitional Rise. The rise of a plume under the influence of the mean wind and the properties of the plume itself; i.e., the rise before atmospheric turbulence or stratification has a significant effect.

Transmissity. The relative lateral movement of ground waters beneath the surface of earth.

Transmissometer. An instrument for measuring the electrical conductivity of a fluid.

Transpiration. The process by which plants dissipate water into the atmosphere from their stomata of leaves or other surfaces.

Tremie Pipe. Common term for well grout pipe.

Trickling Filter. A device for the biological or secondary treatment of waste water consisting of a bed of rocks or stones that support bacterial growth. Sewage is trickled over the bed enabling the bacteria to break down organic wastes.

Trommel. A rotary screen for separating wastes in a resource recovery plant.

Trophogenic Zone. Region of photosynthetic production of organic matter in lakes.

Tropholytic Zone. Region in lakes where breakdown of organic matter occurs.

Tropopause. The region at the top of the troposphere.

Troposphere. The layer of the atmosphere extending seven to ten miles above the earth. Vital to life on earth, it contains clouds and moisture that reach earth as rain or snow.

TRS. Total Reduced Sulfur.

TSCA. Toxic Substances Control Act.

TSM. Transportation System Management.

TSP. Total suspended particulate matter.

TSS. Total suspended solids.

TTHMs. Total trihalomethanes.

Tuberculation. A condition which develops on the interior of pipe lines due to corrosive materials present in water passing through the pipes resulting in formation of lumps which restrict flow.

Tungsten (W). A hard brittle non-magnetic metal element used in filaments and as a hardening alloy in steel. Atomic weight is 184.0 and atomic number is 74.

Turbidimeter. A device used to measure the amount of suspended solids in a liquid.

Turbidity. A thick, hazy condition of air due to the presence of particulates of other pollutants, or the similar cloudy condition in water due to the suspension of silt or finely divided organic matter.

Turbine. An engine that forces a stream of gas or liquid through jets at high pressure against the curved blades of a wheel, thus forcing the blades to turn.

Turbulence. Three-dimensional diffusive motions in a fluid on a macroscopic scale. According to Lumley and Panofsky, turbulence is also rotational, dissipative, non-linear, and stochastic.

Turnaround. The time between shutting down and starting up of process equipment for repair or maintenance.

Turntable. Common name for rotary table used in well construction.

Tuyeres. Openings or nozzles in a metallurgical furnace through which air is blown as part of the extraction or refining process.

TVA. Tennessee Valley Authority.

Twadell Degrees (°TW). Twaddell Degrees is a measure of acid density and strength.

$$\frac{^\circ TW = SP.\ gr.\ (60^\circ/60^\circ F)}{0.005}$$

T/yr. Tons per year.

U

U. See Uranium.

UHF. Ultra high frequency – 300–3000 MH_z.

Ultimate Analysis (of coal). Contains the following, expressed in per cent by weight:

Carbon	(C)	%
Hydrogen	(H_2)	%
Sulfur	(S)	%
Oxygen	(O_2)	%
Nitrogen	(N_2)	%
Moisture	(H_2O)	%
Ash	(H_2O)	%
		100.0%

Ultra Filtration. Waste water treatment employing membranes with extremely fine porosity.

Ultraviolet Ray. Invisible light rays beyond violet in the color spectrum useful in killing bacteria and spores.

ULV. Ultra low volume.

Uncombined Water. Uncombined water is the weight loss which occurs upon drying a sample of the process feed material to constant weight at 105°C.

Underfed Stoker. A stoker consisting of a trough or pot-shaped retort into which coal is forced by an endless screw or ram. Coal is fed to the fire zone by being pushed up from underneath.

Underfire Air. Any forced or induced air, under control as to quantity and direction, that is supplied beneath a grate and passes through a fuel bed.

Unit Operation. Methods by which raw materials may be altered into states, such as vapor, liquid, or solid without being changed into new substances with different properties and composition. Recognition, study, application and control of the principles and factors utilized in a distinct and self-contained process (for example, filtration). This avoids the duplication of effort which attends the study of similar processes as though each process involved a unique set of principles.

Unit Process. Reactions where raw materials undergo chemical change. See unit operation.

Unstable. Possessing hydrostatic instability. An unstable atmosphere has a negative potential temperature gradient.

UOD. Ultimate oxygen demand.

Uranium (U). A radio active hard white metal element with an atomic weight of 238.14 and an atomic number of 92. Used in power production and weaponry and for fluorescent glass and china painting and in photography.

Uranium X_2. See Protoactinium.

Urban Runoff. Storm water from city streets and gutters that usually contains a great deal of litter and organic and bacterial wastes.

Urea Forms. A urea-formaldehyde reaction product that contains more than one molecule of urea per molecule of formaldehyde.

USDA. U.S. Department of Agriculture.

USDC. U.S. Department of Commerce.

USDI. U.S. Department of Interior.

USEPA. U.S. Environmental Protection Agency.

USGS. United States Geological Survey.

UTM. Universal Transverse Mercator System.

V

V. See Vanadium; also an abbreviation for volts(s).

VAC. Volts alternating current.

Vacuum Breaker. A device which relieves a partial vacuum on a water supply line and thus prevents back siphonage from toilet or other plumbing fixtures.

Vacuum Jet (Steam Jet Enjector). A fluid nozzle that discharges a high velocity jet stream across a section chamber that is connected to the equipment to be evacuated. The gas in the chamber is entrained by the jet stream.

Valence. Also called bond or chemical bond; the chemical combining power of an atom. It indicates the number of electrons that can be lost, gained, or shared by an atom in a compound.

Validation. In modelling, a determination of the reliability of the effort by comparing model estimates with measured real data.

Vanadium (V). A gray infusible metal element used as a hardening element in steel, in coloring glass and as a catalyst. Atomic weight is about 50.95 and atomic number is 23.

Vapor. The gaseous phase of substances that normally are either liquids or solids at atmospheric temperature and pressure; for example, steam and phenolic compounds.

Vapor Plume. The stack effluent consisting of flue gas made visible by condensed water droplets or mist.

Vapor Recovery System. System used in petroleum refining for separating a mixed charge of miscellaneous gases and gasolines into desired intermediates for further processing.

Vaporization. The change of a substance from the liquid to the gaseous state. One of three basic contributing factors to air pollution, the others are attrition and combustion.

Variance. Sanction granted by a governing body for delay or exception in the application of a given law, ordinance or regulation.

VC. Vital capacity.

VCM. Vinyl Chloride.

VE. Value engineering; Visible emissions.

Vectofluviometer. A device for separating and measuring the rainfall according to the variation in wind direction during a storm. It consists of a rain gage with a wind vane actuated swinging catchment.

Vector. Disease vector – a carrier, usually an arthropod, that is capable of transmitting a pathogen from one organism to another.

Veering. A term applied to source direction wind shifts in a clockwise pattern.

Venturi Scrubber. A type of high energy scrubber in which the waste gases pass through a tapered restriction (venturi) and impact with low-pressure water. Gas velocities at the restriction are from 15,000 to 20,000 fpm and pressure drops from 10 to 70 inches water gage.

VHF. Very high frequency 3–30 MH_z.

Vi. See Virginium.

VIM. Vehicle Inspection and Maintenance Program (EPA).

Virginium (Vi). The element with the highest equivalent weight. Atomic weight about 224 and atomic number 87.

Virus. A living organism which passes through filters which normally strain out bacteria and which is a disease stimuli whose biological nature is not completely known.

Visbreaking. Viscosity breaking; lowering or "breaking" the viscosity of residual oil by cracking at relatively low temperatures.

Viscosity. The cohesive force existing between particles of a fluid which causes the fluid to offer resistance to flow or being poured.

Visibility. In United States weather observing practice, the greatest distance in a given direction at which it is just possible to see and identify with the unaided eye

(a) in the daytime, a prominent dark object against the sky at the horizon, and (b) at night, a known, preferably unfocused, moderately intense light source. After visibilities have been determined around the entire horizon circle, they are resolved into a single value of prevailing visibility for reporting purpose.

Visible Emission. Means an emission greater than 5 per cent opacity or ¼ Ringelmann measured by standard methods.

VMC. Visual meteorological conditions.

VMT. Vehicle Miles Traveled.

VO. Volatile Organics.

VOC's. Volatile Organic Compounds.

Volatile. Evaporating readily at a relatively low temperature.

Volatile Matter. The gasoline constituents that can be driven off liquids and solids by the application of heat. Specifically for coal, that portion which is driven off in gas or vapor form when coal is subjected to a standardized temperature test.

Volatile Organic Compounds. Any compounds containing carbon and hydrogen or carbon and hydrogen in combination with any other element which has a vapor pressure of 1.5 psia (77.6 mmHg) or greater under actual storage conditions.

V_t. Total volume.

W

W. See Tungsten; also watt.

Waste. Also see solid waste. (1) bulky waste-items whose large size precludes or complicates their handling by normal collection, processing or disposal methods. (2) construction and demolition waste-building materials and rubble resulting from construction, remodeling, repair and demolition operations. (3) hazardous wastes that require special handling to avoid illness or injury to persons or damage to property. (4) special waste – those wastes that require extraordinary management. (5) wood pulp waste – wood or paper fiber residue resulting from a manufacturing process. (6) yard waste – plant clippings, prunings and other discarded material from yards and gardens. Also known as yard rubbish.

Waste Disposal. The orderly process of discarding useless or unwanted material.

Waste Heat Boilers. Boilers which utilize the heat of exhaust gas or process gas to generate steam or to heat water.

Waste Treatment Plant. A series of tanks, screens, filters, and other processes by which most pollutants are removed from water.

Waste Water. Water carrying wastes from homes, businesses and industries that is a mixture of water and dissolved or suspended solids.

Water Crop. See yield.

Water Hardness. A measure of calcium and other salts in water, expressed as mg/liter of calcium carbonate, $CaCO_3$. Hard water causes scale deposits in boiler and water cooling systems and precipitation of salts in food processing.

Water Pollution. The addition of sewage, industrial wastes or other harmful or objectionable material to water in concentrations or in sufficient quantities to result in measurable degradation of water quality.

Water Quality Criteria. The levels of pollutants that affect the suitability of water for a given use. Generally, water use classification includes: public water supply; recreation; propagation of fish and other aquatic life; agricultural use and industrial use.

Water Quality Standard. A plan for water quality management containing four major elements: the use (recreation, drinking water, fish and wildlife propagation, industrial or agricultural) to be made of the water; criteria to protect those uses; implementation plans (for needed industrial-municipal waste treatment improvements) and enforcement plans, and an anti-degradation statement to protect existing high quality waters.

Water Softener. Apparatus designed to remove from water the dissolved calcium and magnesium minerals producing hardness, which creates formation of scale in boilers and hot-water tanks, causing difficulty in making soapsuds, and other undesirable effects.

Water Supply System. The system for the collection, treatment, storage and distribution of potable water from the sources of supply to the consumer.

Water Table. The upper level of ground water.

Watershed. The area contained within a divide above a specified point on a stream that is drained by that stream.

Watt (w). A unit of power equal to the work done at one absolute joule per second or rate of work done by one ampere of electrical current under a pressure of one volt; 1/746 HP.

Waves of Pollution. Successive masses of polluted water occurring at intervals in a moving body of water.

WBC. White blood cells.

Weak Wash. In wood pulping, a liquid stream in the kraft process which results from washing of the lime mud.

Weir. A dam with an overflow of various shapes of crests or notches to divert or change the normal gravity flow of fluids.

Well. An artificial excavation to obtain liquids from the interstices or lens of the subsurface formations which it penetrates.

WERC. Washington Environmental Research Center.

Wet Collectors. Devices which use a variety of methods to wet the contaminant particles in order to remove them from the gas stream (see scrubbers).

Wet Filters. A spray chamber with filter pads composed of glass fibers, knitted wire mesh, or other fibrous materials. The dust is collected on the filter pads.

Wet Well. A compartment to collect drainage or fluids in a system which is so constructed to be capable of being pumped out.

Wetting Agent. A chemical that reduces the surface tension of water and enables it to soak into porous material more readily.

White Liquor. Cooking liquid used in the wood pulping industry. Kraft process: consists of approximately 1/3 sodium sulfide (Na_2S) and 2/3 sodium hydroxide ($NaOH$). Sulfite process: consists of sulfurous acid plus one of the following: calcium bisulfite, sodium bisulfite, magnesium bisulfite, or ammonium bisulfite.

Wind. The natural, horizontal movement of air.

Wind Energy. A secondary form of solar energy that results from the earth's rotation and the heating of the atmosphere and earth's surface by the sun.

Wiresonde. A tethered radio sonde.

WPCF. Water Pollution Control Federation.

WWTP. Waste Water Treatment Plant.

Wx. Abbreviation for weather.

X

Xanthone. A ketone that is the parent of several natural yellow pigments.

X-Axis. The vertical axis of a cartesian coordinate system.

Xe. See Xenon.

Xenodiagnosis. The detection of a parasite by feeding a suitable intermediate host on supposedly infected areas and later examining blood for parasite.

Xenon (Xe). Rarest and heaviest of atmospheric inert gaseous elements. Atomic weight is 131.3 and atomic number is 54.

Xerarch. Developing ecologically in a dry place.

Xeric. Low or deficient in available moisture for the support of life.

Xerophyte. A plant structurally adapted for life and growth with a limited water supply.

X-Radiation. Exposure to x-rays.

Xray. Electromagnetic radiations of extremely short wave length (less than 100 angstroms) that is produced by bombarding a metallic target with fast electrons in vacuum and can penetrate various thicknesses of solids.

X-Scale. Horizontal line; parallel to horizon.

Xylene. Toxic flammable oily isomeric hydrocarbons which are benzene derivatives obtained from wood tar, coal tar and petroleum distillates.

Y

Y. See Yttrium.

Y-Axis. The horizontal axis of a cartesian coordinate system.

Yb. See Ytterbium.

Yield. The quantity of water expressed as a continuous rate of flow or as a volume per unit of time (acre feet per year, etc.) which can be collected for use from surface or groundwater sources in an area or watershed-water crop.

YPS. Yards per sec.

YS. Yield strength (psi).

Y-Scale. Vertical line measurements.

Ytterbium (Yb). A rare mineral element whose atomic weight is 173.04 and atomic number is 70.

Yttrium (Y). A rare earth iron gray metallic element whose atomic weight is 88.92 and atomic number is 39.

Z

Z. Zulu time (Greenwich Mean Time).

Z-Axis. The third axis of a three dimensional coordinate system. It is normal to the plane of the X-Y axis.

ZBB. Zero based budget.

ZD. Zone description (time).

Zenith. The highest point in the celestial sphere directly over the observer.

Zeolite. A chemical compound so imperfectly bound together that its composition will change in accordance with concentration of chemicals in solution in its presence. Zeolites are used in water softening processes.

Zinc (Zn). A bluish white metal element used principally in batteries. Atomic weight is 65.38 and atomic number is 30.

Zing. A shrill humming noise.

Zip. A sudden sharp hissing or sibilant sound; also slang for "nothing" or "zero."

Ziram. A zinc salt used as an agricultural fungicide.

Zirconium (Zr). A silvery gray powder or black powder element used in gas mantles, paints, lacquers, as an abrasive and a compound silicate, zircon is a gem. Atomic weight is 91.22 and atomic number is 40.

ZL. Weather abbreviation for "freezing drizzle."

Zn. See Zinc.

Zoea. An early larval form of decapod crustaceans such as crabs.

Zonate. Marked with or arranged in zones.

Zonule. A little zone.

Zooecology. A branch of ecology dealing with the relation of animals to their environment and to other animals.

Zooflagellate. A flagellate protozoan lacking photosynthesis and other plantlike characteristics.

Zoogenic. Caused by or associated with animals or their activities.

Zoogeography. A branch of biogeography concerned with the geographical distribution animals and especially with the determination of the areas characterized by special groups of animals and the study of the causes and significance of such groups.

Zooid. An entity that resembles but is not wholly the same as a separate individual organism.

Zoology. A science that deals with animals and is the branch of biology concerned with individual and classes.

Zoonosis. A disease communicable from lower animals to man under natural conditions.

Zooparasite. A parasitic animal.

Zoophyte. Invertebrate animals resembling plants such as coral, sea anemone and sponge.

Zooplankton. Planktonic animals that supply food for fish.

ZR. Weather abbreviation for "Freezing rain."

Zr. See Zirconium.

Zyme. Same as enzyme.

Zymurgy. A branch of applied chemistry dealing with fermentation processes.

PART TWO

TABLES AND MONOGRAPHS

TABLE OF CRITERIA AIR POLLUTANTS

POLLUTANT	SYMBOL	DESCRIPTION	HEALTH EFFECTS	FED. AMBIENT STANDARDS
TOTAL SUSPENDED PARTICULATES	TSP	Airborne solid or liquid particulates smaller than 100 microns - dust, fumes, soot, aerosols, oil and grease	Mechanical irritation of eyes, nose, throat linings &lung tissues	260 ug/m^3, max. 24 hr. concentration not to be exceeded more than once/year.
NITROGEN OXIDES	NO_x	Reddish Brown gases which form acids with moisture; in presence of ultra-violet light and hydrocarbon forms PAN and ozone; produced from high temperature combustion of gasoline and other fossil-fuels	Destructive to body tissues expecially in respiratory tract	100 ug/m^3 annual arithmetic mean
SULFUR DIOXIDE	SO_2	Heavy, pungent colorless gas produced primarily from combustion of sulfur-containing fossil fuels	Destructive to lung tissues; drys mucous linings and nose filia	365 ug/m^3, max. 24-hr. concentration not to be exceeded more than once per year.
CARBON MONOXIDE	CO	Colorless, odorless gas resulting from incomplete combustion	Can be fatal in large amounts; headaches, fatigue, confusion in lesser amounts	9 ppm max. 8 hr. concentration not to be exceeded more than once per year.
Photochemical Oxidants (OZONE)	(O_3)	Colorless, pungent gas-ozone formation by photo chemical reaction between solar radiation and nitrogen dioxide in presence of hydrocarbons	Causes eye irritation coughing, choking, headaches and severe fatigue. Lowers body resistance to infection.	80 ppb max.1 hr. concentration not to be exceeded more than once per year
Volatile Organic Compounds (HYDROCARBONS)	VOC (HC)	A series of Free gases containing HC chains in various format	Act synergistically with nitrogen oxides precursor for ozone. See photochemical oxidant effects	160 ug/m^3, max 3 yr. measured 6-9 A.M. not to be exceeded more than once a year

NATIONAL AIR POLLUTION STANDARD INDEX (PSI)*

INDEX VALUE	AIR QUALITY LEVEL	POLLUTANT LEVELS: TSP (24-hour), $\mu g/m^3$	SO_2 (24-hour), $\mu g/m^3$	CO (8-hour), mg/m^3	O_3 (1-hour), $\mu g/m^3$	NO_2 (1-hour), $\mu g/m^3$	HEALTH EFFECT DESCRIPTOR	GENERAL HEALTH EFFECTS	CAUTIONARY STATEMENTS
500	SIGNIFICANT HARM	1000	2620	57.5	1200	3750	HAZARDOUS	Premature death of ill and elderly. Healthy people will experience adverse symptoms that affect their normal activity.	All persons should remain indoors, keeping windows and doors closed. All persons should minimize physical exertion and avoid traffic.
400	EMERGENCY	875	2100	46.0	1000	3000		Premature onset of certain diseases in addition to significant aggravation of symptoms and decreased exercise tolerance in healthy persons.	Elderly and persons with existing diseases should stay indoors and avoid physical exertion. General population should avoid outdoor activity.
300	WARNING	625	1600	34.0	800	2260	VERY UNHEALTHFUL	Significant aggravation of symptoms and decreased exercise tolerance in persons with heart or lung disease, with widespread symptoms in the healthy population.	Elderly and persons with existing heart or lung disease should stay indoors and reduce physical activity.
200	ALERT	375	800	17.0	400[c]	1130	UNHEALTHFUL	Mild aggravation of symptoms in susceptible persons, with irritation symptoms in the healthy population.	Persons with existing heart or respiratory ailments should reduce physical exertion and outdoor activity
100	NAAQS	260	365	10.0	160	[a]	MODERATE		
50	50% OF NAAQS	75[b]	80[b]	5.0	80	[a]	GOOD		
0		0	0	0	0	[a]			

[a]No index values reported at concentration levels below those specified by "Alert Level" criteria.

[b]Annual primary NAAQS.

[c]400 $\mu g/m^3$ was used instead of the O_3 Alert Level of 200 $\mu g/m^3$ (see text).

* EPA-450/2-76-013 Aug. 76, RTP, NC - Guideline for Public Reporting of Daily Air Quality - Pollutant Standards Index (PSI)

AIR POLLUTION PLANNING STATEMENT EPA

All criteria document issuance or reissuance required by Sections 104 and 106 of the 77 Clean Air Act Amendments will conform to the following schedule:

DOCUMENT	START DATE (ON OR BEFORE)	COMPLETION DATE (ON OR BEFORE)
Lead	In progress	12/77
Oxidants	In progress	2/78
NO_2 (Short-Term)	11/77	2/78
NO_x	11/77	1/79
CO	6/78	12/79
SO_x	6/78	5/80
Particulates	6/78	5/80
Hydrocarbons	1/79	9/80

TABLE OF ESTIMATED SEWAGE FLOWS

Type of Establishment	Gallons Per Day (GPD)
COMMERCIAL	
Barber and beauty shops (per chair)	85-100
Bowling alleys (toilet wastes only per lane)	100-200
Country club	
a. per resident member	100-150
b. per non-resident member	25-30
Dentist offices	
a. per dentist	250-275
b. per wet chair	200-225
Doctor offices (per doctor)	250-275
Drive-in movie theatre (per car space)	5-7
Food service operations	
a. Ordinary restaurant (per seat)	50-70
b. 24-hour restaurant (per seat)	75-100
c. Bar and cocktail lounge (per seat)	30-40
d. Drive-in restaurant (per car space)	35-100
Hotels and motels (per bed)	100-125
Laundries, self-service (per machine)	400-450
Office building (per worker)	20-25
Resort camps, cottages (per person)	50-60
Service stations	
a. for first bay	1000-1100
b. for each additional bay	500-600
Shopping centers (without food or laundry) per square foot of floor space	0.2-1.0
Stadiums, race tracks, ball parks (per seat)	3-5
Stores, without food service (per toilet room)	400-450
Theaters, auditoriums (per seat)	3-5
Trailer/mobile home parks	
a. Travel trailer (overnite), w/o water & sewer hookup (per trailer space)	50-60
b. Add for water and sewer hookups, (per trailer space)	50-100
INDUSTRIAL	
Factories, exclusive of industrial wastes (gallons per person per shift)	
a. No showers provided	15-20
b. Showers provided	35-40

(Continued)

TABLE OF ESTIMATED SEWAGE FLOWS (*Continued*)

Type of Establishment	Gallons Per Day (GPD)
Work or construction camps, semi-permanent (per worker)	50-60
INSTITUTIONAL	
Airports	
a. per passenger	5-8
b. per employee	20-25
Churches (per seat)	150-250
Hospitals (per bed)	100-125
Nursing, rest homes (per person)	100-125
Parks, public picnic	
a. with toilets only (per person)	5-8
b. with bathhouse, showers and toilets (per person)	10-15
Public institutions other than hospitals (per person)	75-125
Schools, per student	
a. Day-type	10-15
b. Add for showers	5-8
c. Add for cafeteria	5-8
d. Add for day school workers	15-20
e. Boarding-type	75-100
Swimming pools, public	10-15
RESIDENTIAL	
Residences	
a. Single family (per bedroom)	100-150
b. Apartment (per bedroom)	100-150
c. Mobile home (per unit)	200-300
d. per occupant	100-125

EPA WATER QUALITY CRITERIA TOXIC CATEGORIES*

The 65 Toxic Categories in Consent Degree

Acenaphthene	Copper & compounds	Lead & compounds
Acrolein	Cyanides	Mercury & compounds
Acrylonitrile	DDT & metabolites	Naphthalene
Aldrin/Dieldrin	Dichlorobenzenes	Nickel & compounds
Antimony & compounds	Dichlorobenzidine	Nitrobenzene
Arsenic & Compounds	Dichloroethylenes	Nitrophenols
Asbestos	Dichloropropane	Nitrosamines
Benzene	Dimethylphenol	Pentachlorophenol
Benzidine	Dinitrotoluene	Phenol
Beryllium & Compounds	Diphenylhydrazine	Phthalate esters
Cadmium & Compounds	Endosulfan & Metabolites	PCBs
Carbon Tetrachloride	Endrin & metabolites	Polynuclear hydrocarbons
Chlordane	Ethylbenzene	Selenium & compounds
Chlorinated benzenes	Fluoranthene	Silver & compounds
Chlorinated ethanes	Haloethers	Tetrachlorodibenzo dioxin
Chloroalkyl ethers	Halomethanes	Tetrachloroethylene
Chlorinated naphthalene	Heptachlor & Metabolites	Thallium & compounds
Chlorinated phenols	Hexachlorobutadiene	Toluene
Chloroform	Hexachlorocyclohexane	Toxaphene
Chlorophenol	Hexachlorocyclopentadiene	Vinyl chloride
Chromium & compounds	Isophorone	Zinc & compounds

* EPA to develop effluent limitation standards by December 1979 and 21 affected industries to comply with BAT by 1984.

WATER QUALITY PARAMETERS*

Parameter		Units	Optimal Values	Instrument or Method
Temperature-T		Degrees Celsius	5°–30° C	Thermometer
Dissolved Oxygen-DO		Parts/Million	>5 ppm	Wet Chemistry
Biochemical Oxygen Demand-BOD		Parts/Million	Value which will not depress DO	Wet Chemistry
Turbidity-JU		Jackson Units	<50 JU	Turbidimeter
Ortho-Phosphates		Parts/Million PO_4 as P	<.02	Wet Chemistry
Suspended Solids-55		Parts/Million	<20	Wet Chemistry (Oven)
Fecel Coliform Total Coliform		Most Probable Number	<1000 MPN Daily	Wet Chemistry (Incubator)
Oil and Grease		Parts/Million	<15	Wet Chemistry
Metals				
Iron	Fe	mg/1	<1.0	
Lead	Pb		<.05	Atomic
Zinc	Zn		<1.0	Absorption
Copper	Co		<.5	Spectrophotometer
Cadmium	Cd		.01	
Mercury	Hg		<.005	
Acidity-pH Field		0–15	7.4	ph Meter
Chloride-CL		Parts/Million	Potable 20–250 ppm	Wet Chemistry
Kjeldahl Nitrogen-TKN		Parts/Million	<2	Wet Chemistry
Nitrates		Parts/Million NO_3 as N	<10	Wet Chemistry
Total Solids-TS Dissolved Solids-DS		Parts/Million	<500 <250	Wet Chemistry (Oven)
Carbonaceous Oxygen Demand-COD		Parts/Million	So as not to degrees DO below 5 ppm	Wet Chemistry
Total Substances		Parts/Billion	Various	Various

*Source: Walter W. Honour and Frank L. Cross. *Management of Local Pollution Control Programs,* p122, Technomic, Westport, Conn. 1978.

TEMPERATURE CONVERSION TABLE

°C	°F	°C	°F	°C	°F	°C	F°	C°	F°	C°	°F
-50	-58	-19	-2	10	50	41	106	72	162	103	217
-49	-56	-18	0	11	52	42	108	73	163	104	219
-48	-54	-17	1	12	54	43	109	74	165	105	221
-47	-53	-16	3	13	55	44	111	75	167	106	223
-46	-51	-15	5	14	57	45	113	76	169	107	225
-45	-49	-14	7	15	59	46	115	77	171	108	226
-44	-47	-13	9	16	61	47	117	78	172	109	228
-43	-45	-12	10	17	63	48	118	79	174	110	230
-42	-44	-11	12	18	64	49	120	80	176	115	239
-41	-42	-10	14	19	66	50	122	81	178	120	248
-40	-40	-9	16	20	68	51	124	82	180	125	257
-39	-38	-8	18	21	70	52	126	83	181	130	266
-38	-36	-7	19	22	72	53	127	84	183	135	275
-37	-35	-6	21	23	73	54	129	85	185	140	284
-36	-33	-5	23	24	75	55	131	86	187	150	302
-35	-31	-4	25	25	77	56	133	87	189	160	320
-34	-29	-3	27	26	79	57	135	88	190	170	338
-33	-27	-2	28	27	81	58	136	89	192	180	356
-32	-26	-1	30	28	82	59	138	90	194	190	374
-31	-24	0	32	29	84	60	140	91	196	200	392
-30	-22	1	34	30	86	61	142	92	198	250	482
-29	-20	2	36	31	88	62	144	93	200	300	572
-28	-18	3	37	32	90	63	145	94	201	350	662
-27	-17	4	39	33	91	64	147	95	203	400	752
-26	-15	5	41	34	93	65	149	96	205	450	842
-25	-13	6	43	35	95	66	151	97	207	500	932
-24	-11	7	45	36	97	67	153	98	208	550	1022
-23	- 9	8	46	37	99	68	154	99	210	600	1112
-22	- 8	9	48	38	100	69	156	100	212	650	1202
-21	- 6	10	50	39	102	70	158	101	214	700	1392
-20	- 4	11	52	40	104	71	160	102	216	1000	1832

(Fahrenheit degrees rounded to nearest whole degree except those opposite Celsius degrees which are divisible by 5 or are whole numbers)

TEMPERATURE CONVERSION CHART

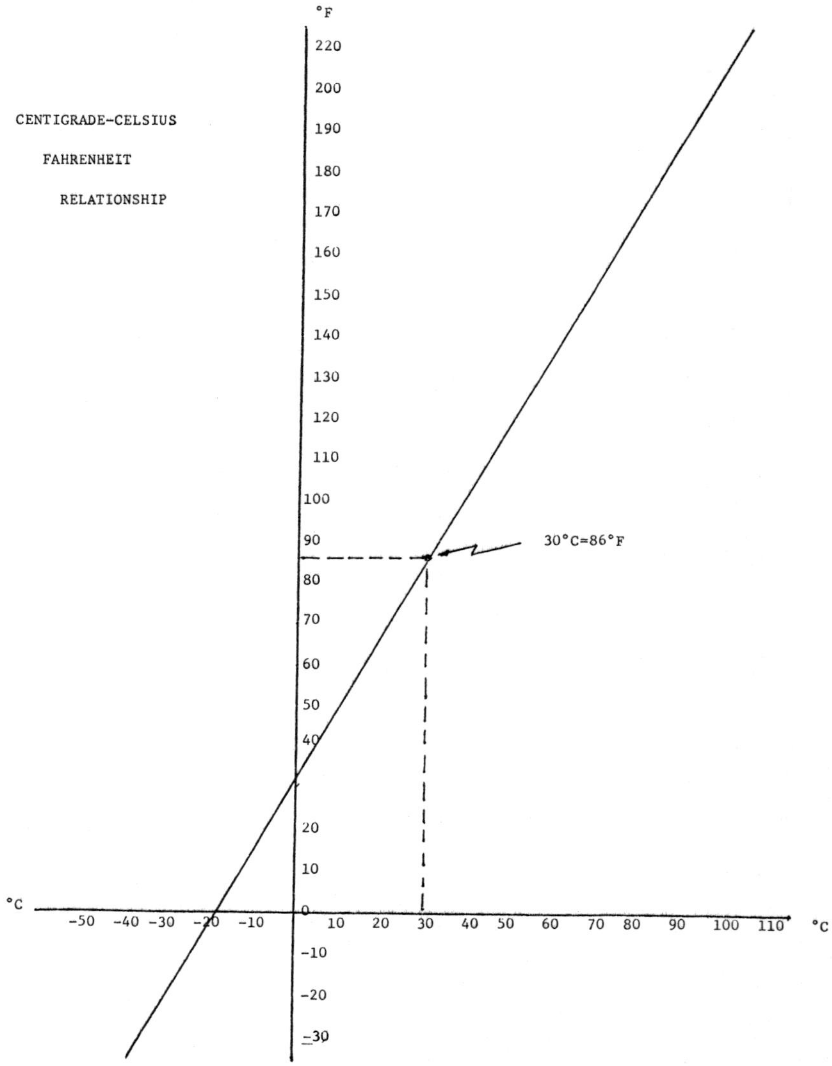

ENERGY EQUIVALENCY TABLE

1 Barrel oil = 6,000,000 BTU's

1 ft^3 Natural Gas = 1,000 BTU's

@ 150 PSI & 200°F, 1,500,000 BTU's req. to make 1,000 lbs steam

1 BTU = 778 ft-lb = 1055 joules = 252 calories

1 erg = 1 dyne-cm

1 joule = 10^7 ergs = 0.738 ft-lb

1 HP = 550 ft-lb/sec = 33,000 ft-lb/min = 2545 BTU/hr = 0.746 Kw

1 HP/hr = 2545 BTU = 0.746 Kwh

1 Kwh = 860 calories (Int) = 1.341 HP/hr

1 watt = 1 joule/sec

1 kilowatt = 1000 watts = 1.341 HP = 3413 BTU/hr.

CONVERSIONS AND FORMULA FOR POLLUTION CONTROL WORK

$$PPM_{(weight)} = \frac{28.8}{(\text{Molecular Weight})} PPM_{(volume)}$$

Commonly Used Molecular Weights

Sulfur Dioxide	SO_2	64
Nitrogen Dioxide	NO_2	46
Ozone	O_3	48
Carbon Monoxide	CO	28
Carbon Dioxide	CO_2	44
Hydrogen Sulfide	H_2S	34

Pressures

1 Atmosphere	=	33.90	Feet of water
	=	29.92	Inches of mercury
	=	14.70	Pounds per square inch
	=	1013.2	Millibars
	=	760.0	Millimeters of mercury
	=	1.058	Tons per square foot
1 Gram per square foot	=	10.764	grams per square meter
	=	96.154	Pounds per acre
	=	30.73	Tons per square mile
1 Micrograms per cubic meter	=	62.43×10^{-12}	pounds per cubic foot
	=	0.02832	Micrograms per cubic foot
	=	(0.02402 ÷ Mol. wgt)	PPM (volume)
	=	834.7×10^{-6}	PPM (weight)
1 pound per square foot	=	6.945×10^{3}	pounds per square inch
	=	0.01394	tons per square mile
	=	0.04356	pounds per acre
	=	4.8824	milligrams per square meter

Velocities

1 foot per second	=	0.3048	meters per second
	=	0.6818	miles per hour (statute)
1 mile per hour (statute)	=	1.467	feet per second
	=	88	feet per minute
	=	0.447	meters per second
	=	26.82	meters per minute
	=	44.70	centimeters per second
	=	1.6093	kilometers per hour
	=	0.8684	knots

CONVERSIONS AND FORMULA FOR POLLUTION CONTROL WORK (*Continued*)

1 foot per minute	=	0.00508	meters per second
1 knot	=	1	nautical mile per hour
	=	1.151	statute miles per hour
1 meter per second	=	3.281	feet per second
	=	3.6	kilometers per hour
	=	2.237	miles per hour
1 cubic foot per minute	=	0.0283	cubic meters per minute

Miscellaneous

1 Micron	=	3.94×10^5	inches
	=	10^{-3}	millimeters
1 cubic foot water	=	61.43	pounds
	=	28.32	kilograms
1 U.S. gallon water	=	8.345	pounds
1 foot pound	=	0.1383	kilogram meters
	=	1.356	joules
1 cubic inch	=	0.0036	Imperial gallons
	=	0.00433	U.S. Gallons
1 cubic foot	=	6.229	Imperial gallons
	=	7.481	U.S. Gallons
1 Imperial gallon	=	1.2001	U.S. Gallons
	=	4.546	Liters
1 U.S. Gallon	=	3.7854	Liters

CONVERSIONS AND FORMULA FOR POLLUTION CONTROL WORK (*Continued*)

Metric – English Measuring Systems Conversions

UNIT	ABBR.	METRIC VALUE	U.S. ENGLISH EQUIV. (APPROX.)
Acre	a	$4047m^2$	$43,560\ ft^2$
Centare (square meter)	ca(m^2)	$1m^2$	$10.76\ ft^2$
Centimeter	cm	15^2m	0.39 in
Cubic centimeter	cc	$10^{-6}m^3$	$0.061\ in^3$
Cubic foot	ft^3	$0.028m^3$	$1728\ in^3$
Cubic inch	in^3	16.387 cc	$0.00058\ ft^3$
Cubic yard	yd^3	$0.765\ m^3$	$27\ ft^3$
Fluid ounce	fl oz	29.573 ml	8 drams
Foot	ft	30.480 cm	12.0 in
Gallon	gal.	3.785 l	4 qts.
Gram	gm	1 gm	0.035 oz
Hectare	ha	10^4m^2	2.47 acres
Hectometer	hm	10^2m	109.36 yds
Hundredweight	cwt	45.359 kg	100 lbs
Inch	in	2.54 om	0.083 ft.
Kilogram	Kg	10^3gm	2.2046 lbs
Kilometer	km	10^3m	0.62 mile
Liter	l	1 l (1000cc)	1.057 qts (liquid)
Meter	m	1 m	39.37 in.
Metric ton	MT	10^6gm	1.1 tons
Mile	mi	1.609 km	5280 ft
Milligram	mg	$10^{-3}gm$	0.015 grains
Milliliter	ml	$10^{-3}l$(1cc)	0.27 drams
Millimeter	mm	$10^{-3}m$	0.04 in
Ounce	Oz	28.349 gm	437.5 grains
Pint	pt		16 fluid oz
Pound	lb	0.453 kg	16 oz.
Quart	qt	0.946 l	2 pts.
Rod	rd	5.029 m	5.50 yds
Square centimeter	cm^2	$10^{-4}m^2$	$0.155\ in^2$
Square foot	ft^2	$0.093\ m^2$	$144\ in^2$
Square inch	in^2	$6.451\ cm^2$	$0.007\ ft^2$
Square Kilometer	km^2	10^6m^2	$0.3861\ mi^2$
Square mile	mi^2	$2.590\ km^2$	640 acres
Square yards	yd^2	$0.836\ m^2$	$9\ ft^2$
Stere	s	$1m^3$	$1.31\ yds^3$
Ton	T	0.907 MT	2000 lbs
Yard	yd	$0.914m^3$	3.0 ft.

YEARLY AVERAGE* EQUIVALENT SOUND LEVELS IDENTIFIED AS REQUISITE TO PROTECT THE PUBLIC HEALTH AND WELFARE WITH AN ADEQUATE MARGIN OF SAFETY

	Measure	Indoor Activity Interference	Indoor Hearing Loss Consideration	To Protect Against Both Effects (b)	Outdoor Activity Interference	Outdoor Hearing Loss Consideration	To Protect Against Both Effects (b)
Residential with Outside Space and Farm Residences	L_{dn}	45		45	55		55
	$L_{eq(24)}$		70			70	
Residential with No Outside Space	L_{dn}	45		45			
	$L_{eq(24)}$		70				
Commercial	$L_{eq(24)}$	(a)	70	70(c)	(a)	70	70(c)
Inside Transportation	$L_{eq(24)}$	(a)	70	(a)			
Industrial	$L_{eq(24)}$(d)	(a)	70	70(c)	(a)	70	70(c)
Hospitals	L_{dn}	45		45	55		55
	$L_{eq(24)}$		70			70	
Educational	$L_{eq(24)}$	45		45	55		55
	$L_{eq(24)}$(d)		70			70	
Recreational Areas	$L_{eq(24)}$	(a)	70	70(c)	(a)	70	70(c)
Farm Land and General Unpopulated Land	$L_{eq(24)}$				(a)	70	70(c)

Code:

a. Since different types of activities appear to be associated with different levels, identification of a maximum level for activity interference may be difficult except in those circumstances where speech communication is a critical activity.
b. Based on lowest level.
c. Based only on hearing loss.
d. An $L_{eq(8)}$ of 75 dB may be identified in these situations so long as the exposure over the remaining 16 hours per day is low enough to result in a negligible contribution to the 24-hour average, i.e., no greater than an L_{eq} of 60 dB.

Note: Explanation of identified level for hearing loss: The exposure period which results in hearing loss at the identified level is a period of 40 years.

*Refers to energy rather than arithmetic averages.

*Source: Environmental Protection Agency, March 1974, Information on Levels of Environmental Noise Requisite to Protect Health and Welfare with an Adequate Margin of Safety.

SOUND PRESSURE LEVELS

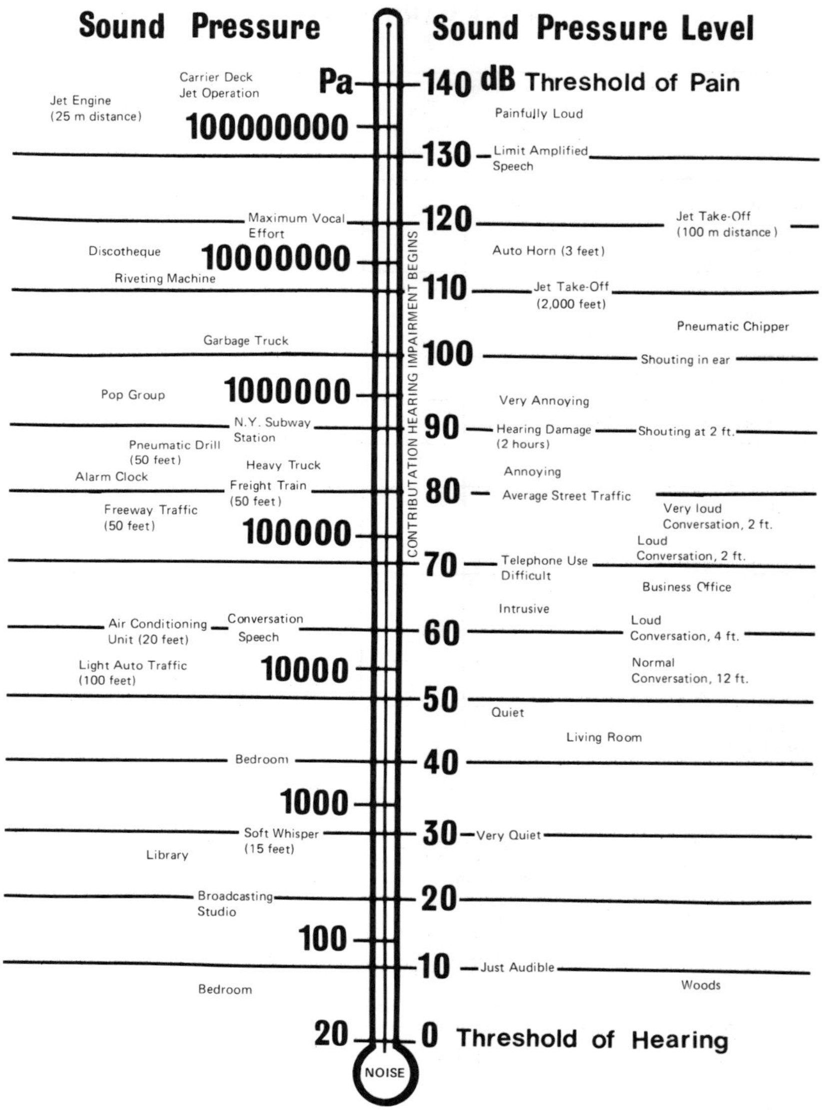

DECIBEL SCALES*

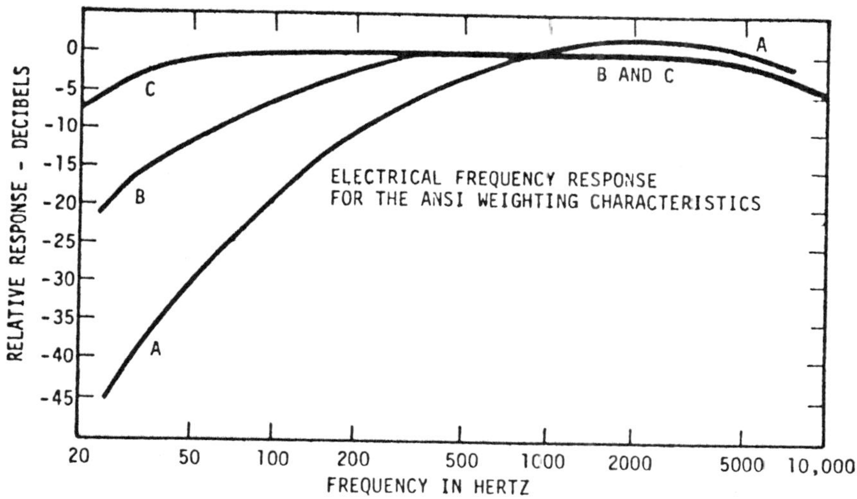

*Source: *A Primer on Noise Measurement,* pamphlet published by General Radio Co. (Concord, Mass. 1972) p6.

TIME

Time in general use throughout the world is basically mean solar time in that a day of 24 hours corresponds closely to one rotation of the earth on its axis and a year of 365 days (366 days in Leap Years) corresponds closely to one revolution of the Earth about the Sun. Although there are other calendars extant, i.e., Chinese, Jewish, and Muhammadan, the above relationship is derived from the Gregorian Calendar. In 46 BC, the Julian Calendar was established in Rome and provided for the numbers of days in the months and for an extra day in February every fourth year. In 1582 AD, however, Pope Gregory XIII, ordered a correction to the Julian Calendar, making a one time correction of the date from October 5th to October 15th and providing in the future that centesimal years would be Leap Years only if divisible by 400. The Gregorian Calendar is used today and has the following numbers of days in each month:

January	31	July	31
February	28	August	31
March	31	September	30
April	30	October	31
May	31	November	30
June	30	December	31

(February has 29 every 4 years on an even numbered year except centesimal years must be divisible by 400)

Each day is divided into 24 hours with the 12 hour period from Midnight to Noon called A.M. (ante meridiem) and the 12 hour period from Noon to Midnight called P.M. (post meridiem). Thus, 6:00 A.M. is six o'clock in the morning and 6:00 P.M. is six o'clock in the evening. There is a growing use internationally of a 24 hour clock system of four digits for each minute of time of the day which has been in use for decades by the military and in marine and aerial navigation. In this system 2:32 A.M. would be written as 0232 and 8:16 P.M. would be written as 2016. Noon would be 1200 and Midnight 2400. One minute after midnight would be 0001.

Each hour is divided into 60 minutes and each minute is divided into 60 seconds. The commonly recognized instant time at any specific location on Earth is determined by the local time zone that includes the location. Earth is divided into 24 local time zones of 15 degrees of longitude each (neglecting some political boundary perturbations). Each local time zone is identified by a Zone Description (ZD). Refer to the figure entitled International Date Line in which the observer is in a point in space looking up at Earth with the South Pole in the center.

All time is reckoned from the mean solar time at the longitudinal meridian (prime meridian) through the Royal Naval Observatory in Greenwich, England. The zero zone description (ZD) time zone including the prime meridian through Greenwich is 7.5 degrees east and west of Greenwich and records the same time and date

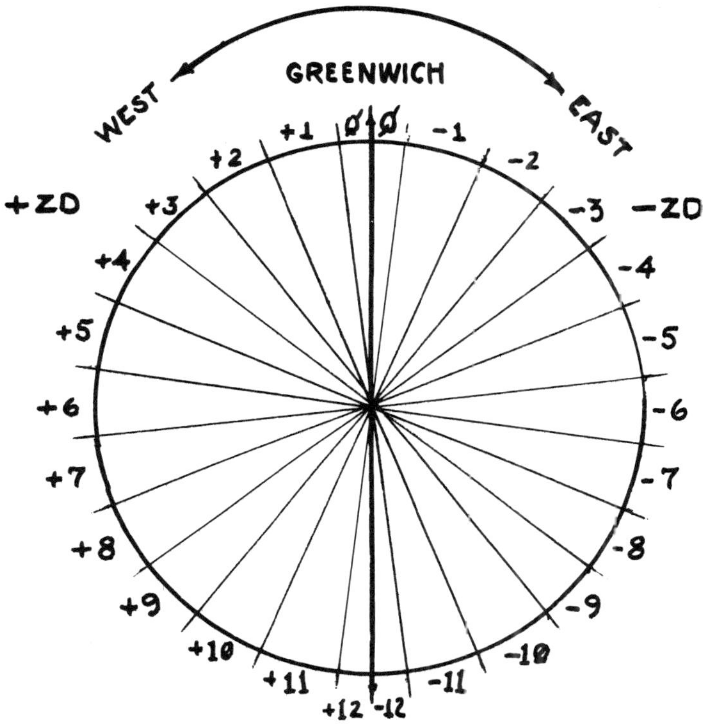

International Date Line 180° E-W λ.

throughout the zone. The time zone 7.5 degrees east and west of the International date line is respectively −12 Zone Description and +12 Zone Description and keeps the same time; however, the date changes when a ship or aircraft crosses the date line. Thus a ship travelling west in the figure above (counterclockwise) and crossing the date line at 0415 on 10 Mar, would not change clock settings but would have a new date of Mar 11.

In all other time zones (neglecting political boundary perturbations) common or local zone time (ZT) throughout the zone corresponds with the central meridian through that zone, i.e., 15°W, 30°W, 45°W etc., and that time can be determined by adding the Zone Description algebraically to the Greenwich Mean Time (GMT). Thus, for New York which is in time zone ZD + 5, if it is 0838 (8:38 A.M.) in Greenwich it is 1338 (1:38 P.M.) in New York). For Moscow which is in time zone ZD-4 it is 0438 (4:38 A.M.) at the same time.

For celestial observations in navigation and determination of monuments in civil engineering local zone time is not adequate for specific locations in a time zone other than those on the central meridian through that zone. It must be remembered that the apparent sun crosses local time zones from East to West and sunrise occurs earlier in the eastern part of the zone. The actual time based on the sun is called Local Civil Time (LCT) which may be defined as: "the time at any specific longitude based upon Greenwich Civil Time (GCT) and the number of degrees minutes, and seconds of longitude the location is East or West of Greenwich, additive if West and subtractive if East." For practical purposes Greenwich Civil Time (GCT) and Greenwich Mean Time (GMT) are the same, the former meaning the specific time at the Prime Meridian while the latter is the time throughout the zero zone description time zone. As an example of the use of LCT, assume that a navigator wishes to establish an accurate position knowing that he is approximately at 65°–15′ Longitude West. In order to make his calculations he must know the local civil time (LCT) of his celestial observations. By maintaining a clock continuously set to Greenwich Civil Time (GCT), he can convert his assumed longitude to time as follows using the conversion table given. Since there is a stotal of 360° of longitude (180E + 180W) about the Earth equating to 24 hours of time then:

Longitude	Time
360°	24 hr.
15°	1 hr.
1°	4 min.
15′	1 min.
1′	4 sec.
15″	1 sec.

The navigator then would convert his Longitude of 65° – 15′W to 4 hr. 21 min. time to be added to GCT to obtain LCT to continue his calculations.

Another use of this conversion table is in the computation of actual sunrise, moonrise or other celestial phenomena for a specific location. If sunrise is given in local zone time and a person who lives a considerable distance to the East of the city wants to know the precise time of sunrise at his residence he can calculate it as follows:

1. Obtain the difference in Longitude in degrees, minutes and seconds between his residence and the center of the city.
2. Convert longitude difference to time using the conversion table.
3. Subtract the time differential from the given sunrise time in the city to obtain the earlier time that sunrise will occur at his residence.

In the above discussions Local Zone Time is the same as Local Standard Time unless Daylight Standard Time is in force in which case the clocks would be advanced one hour in Local Standard Time compared to Local Zone Time.

TIDES AND CURRENTS

Tides and currents are discussed together because there is a relationship between them particularly in estuarine waters.

Tides are caused primarily by the gravitational attraction of the moon and to a much lesser degree by the gravitational attraction of the sun. Other factors affecting the levels and durations of tides are the weather, the hydrography of area of measurement, and seasonal variations stemming from the intervening distances and relative ecliptic positioning of the earth, moon, and sun.

As the Earth rotates on its axis daily the moon's gravitational force pulls surface waters on the Earth on the side facing the moon. A nearly equal and opposite reaction occurs on the opposite side away from the moon causing two high tides approximately 180° of longitude apart. Since the Earth is in constant rotation, the nodal points of these high tides are constantly moving westward following the apparent motion of the moon across the celestial sphere. Also, since there are two high tides always occurring somewhere on the Earth's surface, any specific point on Earth's surface will normally experience two high tides daily. The word "normally" is used because at periodic intervals there may be only one high tide (or low tide) at a specific location in a 24-hour day. This phenomenon occurs primarily because of the lagging of the moon on an average of about 50 minutes a day due to the movement of the Earth in the plane of the ecliptic in its orbit about the sun. It is also caused partially because the plane of the Moon's orbit about the Earth is offset from the equinoctial (plane of the Earth's equator) and the apparent motion of the Moon as viewed from the Earth's surface is not directly east to west. Depending upon Latitude and the season a person at the same location will observe moonrise and moonset occurring about 50 minutes later each day.

The discussion above relating to the diurnal occurrence of high tides also follows for low tides which occur in between the high tides at about six hours from each high tide. The moon's gravitational pull of the surface waters at adjacent and opposite sides of the Earth leave low surface waters at points on the Earth 90° to the high tides nodes.

Levels of tides are affected by strong offshore and onshore winds or wave action. The height of tides in estuarine waters are also affected by rainfall and dry periods.

Twice monthly the Sun, Moon and Earth are approximately in line with the alignment points called syzygy. When the Sun and Moon are on the same side of the Earth they are said to be in conjunction, when on opposite side they are in opposition. Syzygy configuration causes higher than normal tides known as Spring tides. Twice monthly at first and third quarters of the Moon minimum range tides occur which are known as Neap tides.

Since the orbit of the Moon is elliptical, there occurs monthly a closest approach point known as perigee. If perigee should coincide with syzygy, the resultant astronomical condition causes abnormally amplified tides called perigee springs.

The tidal sequence at any measured point consists of low tide, flood, high tide, ebb and repeats continuously. The currents corresponding but not necessarily at the precise time with this sequence are low tide – low slack water, flood – flood current (inbound)high tide – high slack water, ebb current (outbound).

Measurements for shoreline construction, dredge and fill operations, etc. are usually based upon mean high water and mean low water which are the time-weighted averages of high tides levels and low-tides levels.

Tide tables and Tidal Current Tables are published annually by the National Oceanic and Atmospheric Administration for thousands of reference stations around the world in four volumes. For each reference station there are accompanying tables of tidal differences and constants and current variations at inland reference points reckoned from the coastal reference station. By interpolation then, tidal and current information can be calculated for most estuarine and coastal points of interest.

Current is composed of vector and scalar componetns. Set is direction (vector) in which the surface water is flowing.* Drift (scalar) is the amount or rate of current flow expressed in knots (nautical miles per hour).

Upland river, stream and tributary currents are caused by the difference in head and gravity and the amounts are variables depending on hydrography, volume, and angle of fall.

Tidal currents are affected positively and negatively bẏ deltic river current flow adding to or subtracting from tidal action.

Ocean currents are caused by land mass configuration, meteorological conditions, and the coriolis effect of the earth's rotation. There are prevailing currents in geographical areas throughout the world such as the Gulf stream and the Japanese current. Charts are available which depict world ocean currents.

MAP AND CHART COORDINATE SYSTEMS

Maps and charts are used extensively in pollution control work and to a lesser extent are used in energy planning considerations. An understanding of chart development and common coordinate systems is necessary for those who need to locate positions accurately on maps or charts. Generally speaking a map addresses land areas and charts apply to water areas although the word chart is generic and in marine and aerial navigation is used regardless of whether the area is land or water.

Almost all maps and charts in the principal habitated latitudes of the World are based upon Mercator projections. The Mercator projection is developed by envisioning a huge cylinder wrapped around the Earth and tangent at the middle with the Equator. See figure below.

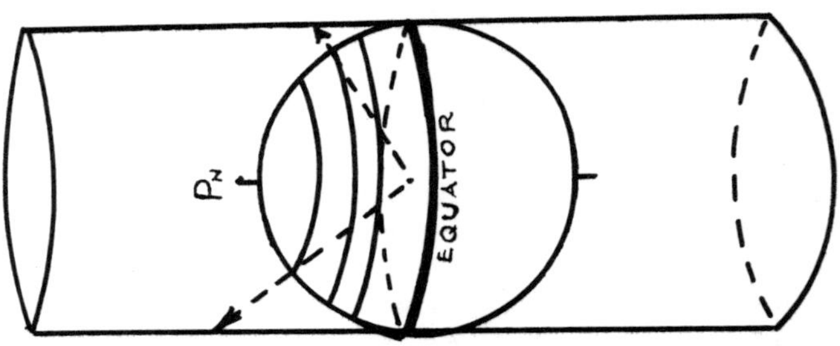

Then envision lines emanating from a source at the center of the Earth through a point on the earth's surface and continuing in space to intersect and describe a similar point on the outer surface of the cylinder. When sufficient points have been projected, the detail of the Earth's surface has been described on the surface of the cylinder. Then visualize the cylinder being cut lengthwise at one point on its

*Expressed in degrees or compass headings.

circumference and being spread out flat to form a rectangle. Portions of this rectangle are then cut up into convenient sections to become maps or charts of local areas of the Earth's surface.

It is obvious that in such a projection, distortion will occur increasing with latitudes North and South of Equator. In fact, the only place there is no distortion is on the Equator. Notwithstanding within specific small areas as depicted by maps and charts, the distortion is unnoticeable. Additionally, some corrective measures are taken in constructing small area maps. There are four types of coordinate systems in common use on maps and charts:

1. Grid system
2. Township platting – Legal Description
3. Universal Transverse Mercator (UTM) system
4. Latitude and Longitude

Grid Systems

Grid systems are encountered on road maps, city maps, and special area maps. On such maps the edges or borders are annotated with evenly spaced consecutive letters or numbers. For example the top and bottom borders might have the letters A through J and the right and left borders the numbers 1 through 10. With such an arrangement the entire map can then be divided into 100 small squares or rectangles. If a legend is provided a person could more easily locate a position by narrowing his scan to one of the smaller squares or rectangles. For example, the township of Wayne might be identified in the legend as in square "B-4," and the user need only to confine his attention to the fourth square down on the left side under "B." In developing point source locations grid overlays on existing maps are frequently used and legends are developed for ease in use of pollution control work.

Township Platting and Legal Description

In pollution control work, it is frequently necessary to determine property ownership and the extent of properties by describing property boundaries.

Most properties in the United States are identified by townships, ranges, sections and fractions of sections. The thirteen original colonies are exceptions to this system and their legal descriptions are more complex since they are a combination of boundaries set by original royal grants, state and local government compromises and the natural geography of the local area involved.

In other than the original thirteen colonies, the U.S. Land Office instigated a government survey based upon rectangular lines coinciding with the meridians of longitude and parallels of latitude. The Land Office designated about 30 principal meridians across the country from which east-west measurements could be made. For example, the north-south meridian at Tallahassee, Florida is a designated principal meridian and is called the Tallahassee Meridian. Through each designated principal meridian is an east-west baseline intersecting usually at or near the geographical point which generated the name of the meridian. Thus an east-west baseline intersects the Tallahassee Meridian at Tallahassee.

North-south sub-meridians parallel to the Principal Meridian are spaced six miles apart East and West and the space between each is called a Range. The first to the West of the principal meridian is called Range 1 West, the second to the West, Range 2 West, the first to the East, Range 1 East and so forth. In a like manner lines parallel to the baseline are drawn running east and west and spaced at six mile intervals North and South of the baseline. The intervals are called townships and they are similarly numbered as they proceed away from the principal baseline. The squares formed by the intersections of the six mile lines are also called Townships, so that then the word refers to the space north and south as well as a specific six by six mile square.

Each square of six by six miles is divided into thirty-six square miles each which are called sections. Sections are numbered starting in the upper right hand corner. See figure below:

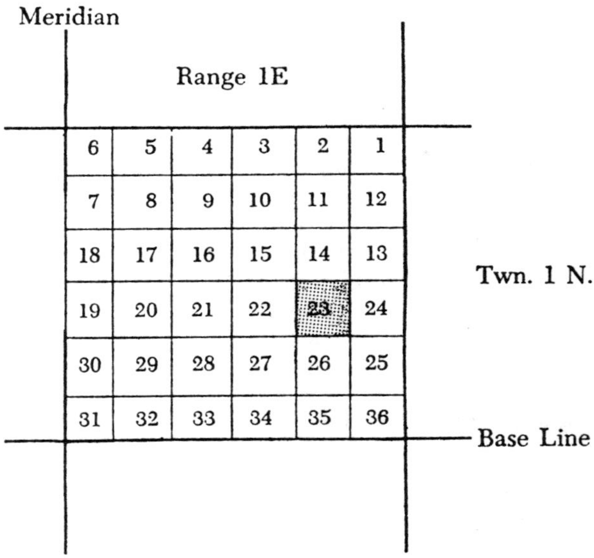

The actual legal description of the shaded area in the Figure would be "all of Section 23, Range 1 East, Township 1 North, Tallahassee Meridian and Baseline Sections are further divided into quarters halfs of quarters, quarters of quarters, quarters of quarters of quarters, and halfs of halfs of quarters. See the following figure.

The legal description of the shaded area in this Figure would be "the SE 1/4 of the SE 1/4 of the SW 1/4 of Section 23, Range 1 East, Township 1 North, Tallahassee Meridian and Base Line."

Platting and plat maps maay further subdivide quarter sections into lot and block numbers and once recorded in the County Public Records Plat Book by subdivision name such as "Lincoln Estates" a legal description can be easily obtained.

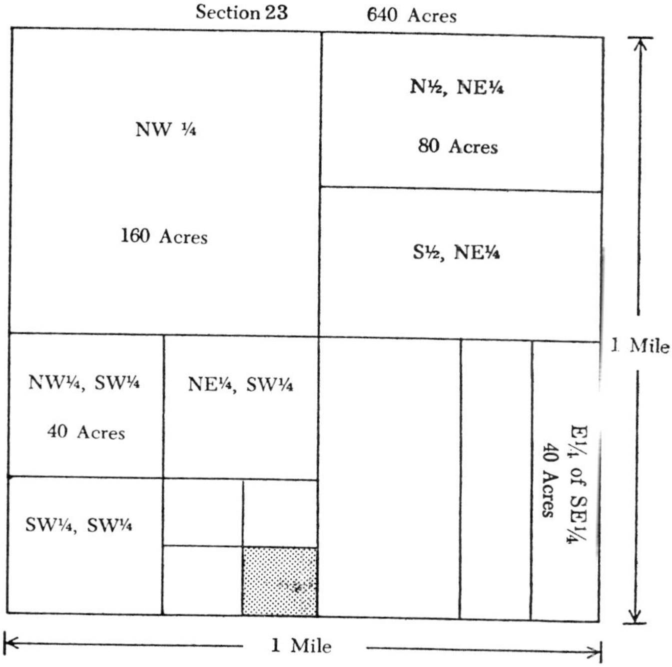

Another method of deriving legal descriptions is based upon monuments, angles and distances and is called metes and bounds. Metes are the measurements and bounds the directions or angles. An example is given "From the Point of Beginning run thence North 65°–19′–22″ East a distance of 412.26 feet to a concrete monument, thence southerly along a curve, concave to the East having a radius of 400.0 feet a distance of 390.6 feet to a concrete monument, thence North 63°–26′–38″ West to the Point of Beginning. Metes and Bounds are used where great accuracy is needed in describing properties and particularly where irregular shapes and patterns are encountered.

Universal Transverse Mercator Projection Systems (UTM)

This system is used by Pollution Control Agencies including the U.S. Environmental Protection Agency for accurate location of point source facilities. U.S. Geodetic Survey (USGS) and Army Map Service (AMS) maps utilize this coordinate system. Basically the UTM system divides the World into 60 zones of 6° of longitude each sequentially numbered starting at the International Date Line, which is the 180° Meridian, East or West Longitude, and numbering to the East. The Western part of the contiguous United States is in Zone 10 and the Eastern most part is in Zone 19. See the figure entitled UTM grid zones.

Alaska is in zones 1 through 09, Hawaii is in zones 04–05, Puerto Rico in zones 19–20, Virgin Islands in zone 20, Guam in zone 55 and American Samoa is in zone 02.

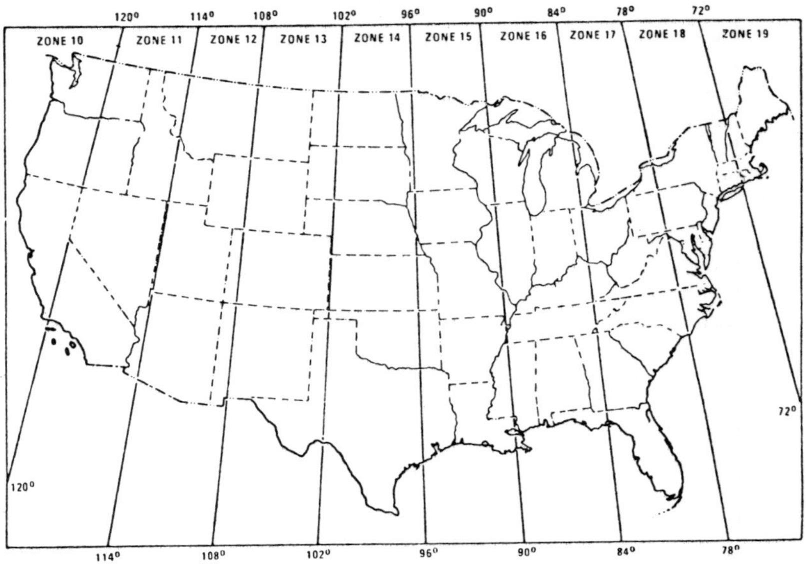

UTM grid zones in the contiguous United States

In all zones the central meridian is 500,000 meters east and since the vertical lines on each side of the 6° zone are 360,000 meters apart, the left vertical line starts at 320,000 meters east and right vertical line ends the zone at 680,000 meters east.

In the northern Hemisphere, the vertical coordinates start at the Equator with zero meters and increase toward the North Pole. In the southern Hemisphere the Equator is listed at 10,000,000 meters north, and numbers decrease in a southerly direction.

UTM maps are available in various scales from the Distribution Section of the U.S. Geological Survey, 1200 S. Eads St., Arlington, Va. 22202:

Scales	Series (0° Lat)
1 24000	7.5 minute
1:62500	15 minute
1:125,000	30 minute
1:250,000	1° X 2°

There is an index map for each state which shows by overlays the maps available and the smallest scale maps usually chosen for environmental work.

On each map the UTM numbers and latitude and longitude will be indicated in the borders with the former in four digit numbers of meters (e.g. 3863m) and the latter in degrees, minutes, and seconds (e.g. 30°22′ 16″). The UTM divisions are marked by blue tick marks in the border.

Individual positions of points on the UTM map are determined by starting at the left border and interpolating the meter number corresponding to an imaginary vertical line through the position to obtain the east coordinate. The north coordinate is obtained similarly by starting at the bottom border and interpolating a meter number for a horizontal line through the position. The meter numbers on the top and bottom borders are 3 digits each and an East coordinate is recorded by putting the zone number before the three numbers and the interpolation in thousands (three numbers) following the nearest number read. Thus a position slightly to the right of 439 in Zone 17 would be recorded as E-17439200. The meter numbers on the left and right borders are four digits each and a North coordinate is recorded by placing the interpolation in thousands (3 numbers) after the nearest number read thus, a position slightly north of 3359 would be recorded as N-3359100. For the examples given a full rendition of UTM coordinates in the proper order would be:

UTM: E-17439500, N-3359100

Latitude and Longitude

Latitude and longitude abbreviated, as (Lat., Long.)or (L, λ) or sometimes (L, Lo) has been in use for describing positions on the Earth's surface and on charts for centuries. Latitude is measured from 0° at the Equator to 90° North at the North Pole and 90° South at the South Pole. Each degree of Latitude is divided into 60 minutes, and the minutes are further subdivided into 60 seconds each. Sometimes for precise monument positioning to the nearest foot seconds of latitude are divided into hundreds. A proper notation of latitude includes the abbreviation for latitude, the degrees, a dash, the munutes, a dash, the seconds, and the designation North or South:

L30–22′–16.20″ N

The nautical mile is derived from the approximate circular distance from the Equator to the North Pole. That distance is assumed to be 5400 nautical miles by designation and because there are (90° X 60′) minutes of Latitude from the Equator to the Pole. Thus, each minute of Latitude is equal to one nautical mile. The significance of this deliberate arrangement devised by the scientific advisers to Napoleon I, is that on all small area charts a scale of nautical miles is unnecessary. The user of such a chart can, with dividers, pick off any desired distance in nautical miles by using the vertical latitude minute scales on each side of the chart. Care must be exercised to use the adjacent minutes of latitude because of the Mercator projection distortion previously described.

The Longitude coordinates were briefly described in the section on Time. Basically, the meridians of longitude are semi-circles of innumerable great circles on the surface of the Earth running through the North and South Poles. Longitude is measured East and West from the Prime Meridian through Greenwich, England which is zero degrees longitude. The 360° of the Earth's surface is divided into

180° East and 180° West longitude. Like latitude, longitude is divided into 60 minutes for each degree, and 60 seconds for each minute and sometimes for very precise positioning into hundreds of seconds. A proper recording of longitude includes the abbreviation, the degrees, a dash, the minutes, a dash, the seconds, and the designation of East or West.

$$L_o\ 79°–26'–31.02''\ W$$

Most commonly in pollution control work, particularly in water quality control, point sources are recorded in latitude and longitude to the nearest second. Such a recording is exemplified in the proper manner as follows:

$$L32°–16'–12''N;\ L_o\ 84°–27'–35''\ W$$

It should be noted that latitude is given first which is the opposite of the UTM system where the Easterly meridianal coordinate is listed first.

BIBLIOGRAPHY

A Floridians Guide to Solar Energy, State Energy, University of Florida, Florida Solar Energy Center, Cape Canaveral, 1975.

Air Monitoring Strategy for State Implementation Plans, Research Triangle Park, N.C., June 1977.

An Ecology Glossary, Reprint from Oil Gas and Petrochemical Equipment Sept-Dec 1974 by the Foxboro Co., Foxboro, Mass, 1975.

Common Environmental Terms, G. J. Stoddard USEPA, Washington, D.C. 1973.

Field Operations and Enforcement for Air Pollution Control, Vol. II, USEPA Washington, D.C. 1972.

Glossary – Water and Sewage Control Engineering, American Society of Civil Engineers, New York, N.Y. 1949.

Ground Water and Wells by Johnson Division, Universal Oil Products Co., St. Paul, Minn 1976.

Guide for Compiling a Compreshensive Inventory, USEPA Washington, D.C. 1973.

Handbook of Chemistry and Physics (22nd Ed), Chemical Rubber Publishing Co., Edited by Charles D. Hodgeman, 1937.

Handbook of Engineering Fundamentals, O. W. Eschbach, J. Wiley & Sons, New York, N.Y. 1936.

Health Effects for Short Term Exposure to Nitrogen Dioxide, USEPA, Washington, D.C. Dec 1977.

Information on Levels of Environmental Noise Requisite to Protect Health and Welfare with an Adequate Margin of Safety, USEPA, Mar 1974.

Interim Guidelines on Air Quality Models, USEPA, Research Triangle Park, N.C. Oct 1977.

Real Estate and Florida Licensing Law, 5th Ed., B-J Publishing Co., Gainesville, Fla 1977.

Technical Guidance for Control of Industrial Process Fugitive Particulate Emissions, USEPA, Research Triangle Park, N.C. Mar 1977.

WASHMIC, Honour, W. W. & Kossan, R. L., Rany Co., Washington, D.C. 1971.